"This timely book is required reading for all scholars of the international climate change negotiations. With a welcome focus on developing countries, it throws light on the varying dynamics of different coalitions and the critical – often unrecognized – roles they have played at key moments in the climate change process. Both newcomers to the climate change negotiations, and longstanding observers, have much to learn from this important volume."

Joanna Depledge, Editor of Climate Policy Journal, *UK*

"Coalitions are a pervasive feature of multilateralism. States use them to increase their bargaining power and reduce the complexity of international negotiations, and yet we know surprisingly little of them. This volume fills this gap, by providing a rich collection of in-depth case studies and conceptual work on coalition formation, maintenance, and effectiveness in the international climate change regime. It is an essential read for scholars and students of international relations interested in the role of coalitions in the international climate change regime and beyond."

Stefan Aykut, University of Hamburg, Germany

"There has been a major gap in the academic literature on the role, formation, and operation of coalitions in multilateral negotiations. This volume responds to this research gap by examining coalition dynamics in the climate regime. The complex economic and environmental nature of the climate regime has led to the development of a plethora of shifting coalitions that shape and are shaped by the negotiating dynamics, whether negotiating the Paris Agreement itself or the Paris Rulebook or the second commitment period of the Kyoto Protocol. This volume not only analyses the different and overlapping coalitions in the climate regime, but it also expands the literature on coalitions and their role in multilateral negotiations."

Pam Chasek, Manhattan College, USA

Coalitions in the Climate Change Negotiations

This edited volume provides both a broad overview of cooperation patterns in the UNFCCC climate change negotiations and an in-depth analysis of specific coalitions and their relations.

Over the course of three parts, this book maps out and takes stock of patterns of cooperation in the climate change negotiations since their inception in 1995. In Part I, the authors focus on the evolution of coalitions over time, examining why these emerged and how they function. Part II drills deeper into a set of coalitions, particularly "new" political groups that have emerged in the last rounds of negotiations around the Copenhagen Accord and the Paris Agreement. Finally, Part III explores common themes and open questions in coalition research, and provides a comprehensive overview of coalitions in the climate change negotiations.

By taking a broad approach to the study of coalitions in the climate change negotiations, this volume is an essential reference source for researchers, students, and negotiators with an interest in the dynamics of climate negotiations.

Carola Klöck is an Assistant Professor at Sciences Po Paris, France.

Paula Castro is a Research Associate at the Zurich University of Applied Sciences, Switzerland.

Florian Weiler is Assistant Professor at the Central European University, School of Public Policy, Vienna, Austria.

Lau Øfjord Blaxekjær is Affiliated Researcher at the University of the Faroe Islands.

Routledge Research in Global Environmental Governance

Global environmental governance has been a prime concern of policymakers since the United Nations Conference on the Human Environment in 1972. Yet, despite more than 900 multilateral environmental treaties coming into force over the past 40 years and numerous public–private and private initiatives to mitigate global change, human-induced environmental degradation is reaching alarming levels. Scientists see compelling evidence that the entire Earth system now operates well outside safe boundaries and at rates that accelerate. The urgent challenge from a social science perspective is how to organize the co-evolution of societies and their surrounding environment; in other words, how to develop effective and equitable governance solutions for today's global problems.

Against this background, the *Routledge Research in Global Environmental Governance* series delivers cutting-edge research on the most vibrant and relevant themes within the academic field of global environmental governance.

Series Editors:
Philipp Pattberg, VU University Amsterdam and the Amsterdam Global Change Institute (AGCI), the Netherlands.
Agni Kalfagianni, Utrecht University, the Netherlands.

Traditions and Trends in Global Environmental Politics
International Relations and the Earth
Edited by Olaf Corry and Hayley Stevenson

Regime Interaction and Climate Change
The case of international aviation and maritime transport
Beatriz Martinez Romera

The Anthropocene Debate and Political Science
Edited by Thomas Hickmann, Lena Partzsch, Philip Pattberg, and Sabine Weiland

Coalitions in the Climate Change Negotiations
Edited by Carola Klöck, Paula Castro, Florian Weiler and Lau Øfjord Blaxekjær

For more information about this series, please visit: https://www.routledge.com/ Routledge-Research-in-Global-Environmental-Governance/book-series/RRGEG

Coalitions in the Climate Change Negotiations

Edited by Carola Klöck, Paula Castro, Florian Weiler, and Lau Øfjord Blaxekjær

Routledge
Taylor & Francis Group

LONDON AND NEW YORK

First published 2021
by Routledge
2 Park Square, Milton Park, Abingdon, Oxon OX14 4RN

and by Routledge
52 Vanderbilt Avenue, New York, NY 10017

Routledge is an imprint of the Taylor & Francis Group, an informa business

British Library Cataloguing-in-Publication Data
A catalogue record for this book is available from the British Library

Library of Congress Cataloging-in-Publication Data
A catalog record has been requested for this book

ISBN: 978-0-367-31321-0 (hbk)
ISBN: 978-0-429-31625-8 (ebk)

Typeset in Times New Roman
by Deanta Global Publishing Services, Chennai, India

Contents

Figures

Tables

Contributing authors

Lau Øfjord Blaxekjær is an Affiliated Researcher at the University of the Faroe Islands, where he directed the international master's programme in West Nordic Studies, Governance and Sustainable Management from 2014 to 2018. Since 2019, Lau has been Special Advisor in the Danish Maritime Authority, leading the work on climate action and climate negotiations in the International Maritime Organization. Lau holds a PhD degree in Political Science from the University of Copenhagen. His research focuses on global climate governance and diplomacy and green growth, as well as sustainability and entrepreneurship in the Arctic.

George Carter is a Research Fellow in Geopolitics and Regionalism at the Australian National University. His research interests explore small island states' influence on decision-making processes in multilateral climate change negotiations. Furthermore, he teaches and researches Pacific states' approaches in foreign policy-making, development, climate and human security, and Pacific Studies.

Paula Castro is a research associate at the Center for Energy and Environment of the Zurich University of Applied Sciences. Her work analyses the role of institutions and bargaining strategies in the climate negotiations, and the adoption and effectiveness of policy instruments for climate change mitigation. Paula has held postdoctoral positions at the University of Zurich (where she also obtained her PhD) and the University of Duisburg-Essen.

Nicholas Chan is lecturer in global studies at the School of Arts and Social Sciences, Monash University Malaysia. Nicholas works on developing country participation in global environmental politics, with a focus on the climate change negotiations and global ocean governance. He has served as pro bono advisor to various small island state delegations to most climate COPs between 2011–2018. Nicholas holds a PhD from the University of Oxford.

Simon Chin-Yee is a research associate at King's College London's Department of War Studies and a teaching fellow in the University College London's School of Public Policy. Simon's work focuses on human security and climate justice with a particular emphasis on vulnerable populations. Holding a PhD from the Politics Department at the University of Manchester, Simon also has extensive experience in international cooperation and public policy, having worked on multiple UN projects primarily in Africa.

Fang Fang is Operations Director of Nature Resources Defence Council's Beijing Office. Before that, she served as Vice President for the Innovation Center for Energy and Transportation (*i*CET), where she implemented the China Energy and Climate Registry project. Prior to that, she worked as a project officer in the International Department, Ministry of Finance, PR China. Fang obtained her MPhil in Environmental and Development Economics from the University of Oslo in Norway.

Lucia Green-Weiskel is an adjunct professor of Political Science at Northern Vermont University-Johnson. She is also a Special Advisor at the Innovation Center for Energy and Transportation (*i*CET), an independent climate change policy centre based in Beijing. She holds a PhD in Political Science from the Graduate Center at the City University of New York. Her doctoral research focuses on US–China cooperation on climate change. Lucia was the manager of the China Energy and Climate Registry, the first online tool for carbon footprint calculation and management developed in China.

Carola Klöck is assistant professor of political science at the Centre for International Research (CERI) at Sciences Po Paris. Carola works on climate change negotiations as well as local adaptation to climate change, both with a focus on small island states. Carola holds a PhD from ETH Zurich (Switzerland) and held positions at the Universities of Gothenburg (Sweden), Antwerp (Belgium), and Göttingen (Germany) before joining Sciences Po.

Bård Lahn is a researcher at CICERO Center for International Climate Research, and a PhD candidate in Science and Technology Studies at the University of Oslo. His work focuses on the relationship between climate science and policy, North/South dimensions of climate justice, and the UNFCCC negotiations. He has previously worked with a number of NGOs and activist networks, and participated in Norway's delegation to the UNFCCC from 2008 to 2011.

Tobias Dan Nielsen is a policy expert at IVL Swedish Environmental Research Institute. Tobias works on environmental and climate policy, and has done extensive research on climate change negotiations, including on political groups and forestry. He has attended six UNFCCC COPs and intersessional meetings. Tobias holds a PhD from the Department of Political Science at Lund University (Sweden).

Joshua Watts has studied and researched climate and energy politics at the University of Cambridge and Erasmus University Rotterdam. His research interests are regional politics of climate change and energy security in Europe and Latin America.

Florian Weiler is Assistant Professor at the Central European University, School of Public Policy. Florian holds a PhD from ETH Zurich, where he worked on the global political economy of climate change. Since finishing his dissertation, he held academic positions at the University of Bamberg (Postdoc), the University of Kent (Lecturer), the University of Basel (Senior Researcher), and the University of Groningen (Assistant Professor). His recent academic work focuses on environmental issues on the one hand, and on stakeholder strategies to influence policy processes on the other.

Acknowledgements

As with any book, this volume is the result of the work of a lot of different people. We want to use this opportunity to acknowledge the support and input of all of them, even if listing only some by name.

This book emerged after we put together a conference panel on "Coalitions, cooperation and conflict in the climate change negotiations" for the 2018 Earth System Governance conference in Utrecht. Some of the papers presented there made it into the present volume, in revised form. We would like to thank all panellists and participants for their input, questions, and for encouraging us to extend our work on coalitions and cooperation. Lau Øfjord Blaxekjær was then the first to suggest an edited volume.

When we approached Routledge with this idea, we were met with enthusiasm, and we would like to sincerely thank the entire team at Routledge Environment and Sustainability, and in particular Annabelle Harris and Matthew Shobbrook for their support, encouragement, and patience. From the first idea of an edited volume to the compilation of the manuscript, Annabelle Harris and Matthew Shobbrook were extremely helpful and understanding, and accompanied us in the writing and rewriting of the book, despite things taking considerably longer than expected. We would also like to thank the three anonymous reviewers of the book proposal for their helpful feedback to further improve and flesh out this volume.

Finally, we would like to express our gratitude – in the name of all contributing authors – to the many negotiators and delegates at the various climate summits who shared their perspectives and experiences with us researchers and who made it possible for us to study and better understand how the climate change negotiations work, and what role coalitions play in them.

Acronyms

Acronym	In full
ABU	Argentina – Brazil – Uruguay
ACF	Advocacy Coalition Framework
ACPC	African Climate Policy Centre
ADP	Ad Hoc Working Group on the Durban Platform for Enhanced Action
AfDB	African Development Bank
AGN	African Group of Climate Change Negotiators
AGW	Africa Group of the Whole
AILAC	Independent Association of Latin America and the Caribbean
AIMS	Africa-Indian Ocean-Mediterranean-South China Sea (SIDS subregion)
ALBA	Bolivarian Alliance for the Peoples of our America
ALBA-TCP	Bolivarian Alliance for the Peoples of our America – People's Trade Treaty
AMCEM	African Ministers Conference of Environment Ministers
AOSIS	Alliance of Small Island States
AU	African Union
AUA	African Union Assembly
BASIC	Brazil – South Africa – India – China
CACAM	Central Asia, Caucasus, Albania, and Moldova Group
CAHOSCC	Committee of African Heads of State and Government on Climate Change
CARICOM	Caribbean Community
CBDR-RC	Common but differentiated responsibilities and respective capabilities
CDM	Clean Development Mechanism
CDPA	Cartagena Dialogue for Progressive Action; also CD, Cartagena
CfRN	Coalition for Rainforest Nations
CG-11	Central Group 11
COMIFAC	Commission des Forêts d'Afrique Centrale
COP	Conference of the Parties
CVF	Climate Vulnerable Forum
DA	Durban Alliance
DG CLIMA	EU's Directorate-General for Climate Action
EIG	Environmental Integrity Group
ENB	Earth Negotiations Bulletin
EU	European Union

FAO	Food and Agriculture Organization of the United Nations
FIELD	Foundation for International Environmental Law and Development
G77	Group of 77 and China; also G77 & China
GCF	Green Climate Fund
GDP	Gross Domestic Product
GRULAC	Group of Latin America and Caribbean Countries
HAC	High Ambition Coalition
HI	Historical institutionalism
IISD	International Institute for Sustainable Development
INC	Intergovernmental Negotiating Committee
INDC	Intended Nationally Determined Contributions
IPCC	Intergovernmental Panel on Climate Change
JUSSCANNZ	Japan, United States, Switzerland, Canada, Australia, Norway, New Zealand
KP	Kyoto Protocol
LAS	League of Arab States
LDCs	Least Developed Countries
LMDCs	Like-Minded Developing Countries
MLDCs	Mountainous Landlocked Developing Countries
MRV	Monitoring, reporting, verification
NAP	National Adaptation Plan
NAPA	National Adaptation Plan of Action
NDCs	Nationally Determined Contributions
NGO	Non-governmental organisation
OAPEC	Organization of Arab Petroleum Exporting Countries
OAU	Organisation of African Unity
OECD	Organisation for Economic Cooperation and Development
OPEC	Organization of Petroleum Exporting Countries
PACJA	Pan-African Climate Justice Alliance
RDT	Resource Dependence Theory
REDD	Reducing Emissions from Deforestation and Forest Degradation
SB	Subsidiary Bodies (to the UNFCCC)
SBI	Subsidiary Body for Implementation
SBSTA	Subsidiary Body for Scientific and Technological Advice
SICA	*Sistema de Integración Centroamericana,* Central American Integration System
SIDS	Small Island Developing States
SPREP	Secretariat of the Pacific Regional Environment Program
UN	United Nations
UNECA	United Nations Economic Commission for Africa
UNEP	United Nations Environment Programme
UNFCCC	United Nations Framework Convention on Climate Change
UNGA	United Nations General Assembly
USA	United States of America
V20	Vulnerable 20
WCC	World Climate Conference
WTO	World Trade Organization

1   Introduction

*Carola Klöck, Paula Castro, Florian Weiler,
and Lau Øfjord Blaxekjær*

Introduction

Climate change is undoubtedly one of the great challenges of the 21st century. For
about 30 years, global and local communities have sought to tackle this challenge.
In 1988, the United General Assembly (UNGA), at the initiative of Malta, recognised climate change as "a common concern of mankind" (United Nations, 1988),
and the United Nations Environment Programme and the World Meteorological
Organisation were tasked with establishing the Intergovernmental Panel on
Climate Change (IPCC). Climate change entered the international agenda, and
has only increased in importance since.

Multilateral negotiations on a global climate agreement started in 1990 – also
the year in which the IPCC published its first assessment report. Only two years
later, negotiations culminated in the United Nations Framework Convention on
Climate Change (UNFCCC). Since 1995 – one year after the Convention's entry
into force – the international community has met annually at the Conference of
the Parties (COP) to further negotiate and implement the Convention's ultimate
objective of "prevent[ing] dangerous anthropogenic interference with the climate system" (UNFCCC, 1992). These climate summits have become "environmental mega-conferences" (Gaventa, 2010) with thousands of participants, and
receive considerable academic, political, media, and public attention (Bäckstrand,
Kuyper, Linnér, & Lövbrand, 2017; Lövbrand, Hjerpe, & Linnér, 2017; Schmidt,
Ivanova, & Schäfer, 2013; Schroeder & Lovell, 2012).

A core feature of the climate change negotiations, and in fact of any multilateral negotiation, is that many states do not negotiate individually, but through
groups or coalitions; Dupont (1996) even defines "negotiations as coalition building". At the same time, some states, particularly larger ones, may also engage in
negotiations individually, although they typically are also members of a coalition.
In the context of multilateral negotiations, coalitions can be defined as cooperative efforts between at least two parties to obtain common goals (e.g. Elgström,
Bjurulf, Johansson, & Sannerstedt, 2001; Narlikar, 2003; Starkey, Boyer, &
Wilkenfeld, 2005). For some, this refers only to short-range, issue-specific objectives (Dupont, 1994, p. 148; Gamson, 1961). For us, coalitions are more long-
term, and refer to repeated coordination to obtain shared objectives.

The terminology used in research and practice differs widely, ranging from alliances, negotiating groups, climate clubs, dialogues, or blocs. Some scholars use these terms to refer to specific types of coalitions. Narlikar (2003), for example, distinguishes issue-specific strategic "alliance coalitions" from broader, ideology-based "bloc coalitions". In line with this divergent terminology, coalitions in the climate negotiations have divergent names such as "*Alliance* of Small States", "*Coalition* of Rainforest Nations", "Environmental Integrity *Group*", "The Cartagena *Dialogue* for Progressive Action", or "*Association* of Independent Latin American and Caribbean Countries" (our emphasis). For simplicity, in this volume we refer to coalitions throughout, regardless of the specific form of that coordination. Coalitions can thus be broad or issue-specific, strategic or ideology-based, ad-hoc or long-term, etc. Indeed, one of the objectives of the present volume is to develop some way of differentiating between coalition *types*. Before we proceed to outline the individual contributions of this volume, let us briefly review coalition research, in particular work on coalitions in multilateral climate negotiations.

Coalition research in and beyond climate change negotiations

Coalitions exist in multiple contexts beyond multilateral negotiations, and research has paid far more attention to coalitions outside of multilateral negotiations. A large part of coalition research is interested in coalition building from a game theory perspective (Bandyopadhyay & Chatterjee, 2012; Gamson, 1961). Much of this research remains conceptual. Applications to climate change are rare and remain rather theoretical (e.g. Buchner & Carraro, 2006; Wu & Thill, 2018). More common are applications to multi-party coalitions in parliamentary democracies and bargaining in business and organisational studies (e.g. Agndal, 2007; Stevenson, 1985), as well as individual behaviour in psychology (e.g. Bazermann, Curhan, Moore, & Valley, 2000). Many of these studies are also theoretical, or based on experiments, often with student subjects, and focus on small-scale negotiation settings with just two or three actors (e.g. Sagi & Diermeier, 2017). In international relations, coalition-building in the European Union has been studied in some depth (Bailer, 2004; Elgström et al., 2001; Finke, 2012), yet because of their majority voting system, these negotiations resemble government coalition-building at the national level, rather than multilateral UN negotiations.

Multilateral UN negotiations, such as those on climate change, are very different and much more complex: they involve almost 200 parties; they cover a large agenda with multiple, highly technical, and partly overlapping items; they are long-term with regular, repeated interactions; and they work through consensus, rather than majority voting. The intricacies of multilateral negotiations are less studied and less well-understood than bilateral settings, and thus require additional work (Crump & Zartman, 2003; Gray, 2011).

In theory, coalitions fulfil two essential functions (Dupont, 1994, 1996): they reduce the complexity of multilateral negotiations, and they increase members' negotiation capacity and bargaining power.

To come together in a coalition, states must share some objectives and positions (Atela, Quinn, Arhin, Duguma, & Mbeva, 2017; Bhandary, 2017; Ciplet, Khan, & Roberts, 2015; Costantini, Crescenzi, Filippis, & Salvatici, 2007). By highlighting commonalities in state positions, and by reducing the number of actors and positions, coalitions reduce complexity and make the process more manageable (Dupont, 1996). Additionally, coalitions improve members' negotiation capacity by allowing them to pool resources and information, and they are thus able to engage more effectively in negotiations. Finally, coalitions increase members' bargaining power, as positions shared by several states carry more weight than those of individual states (Dupont, 1996; Rubin & Zartman, 2000). These functions make coalitions particularly relevant for smaller and less powerful countries (Chasek, 2005; Narlikar, 2003; Penetrante, 2013; Williams, 2005).

On the other hand, coalitions also add a layer to the negotiations, and therefore represent "negotiation within negotiation" (Starkey et al., 2005, p. 40). This additional layer does not come without costs. Coalition formation and maintenance require significant coordination efforts, which represents a challenge in particular for smaller and poorer countries who mostly can send only small delegations to negotiations (Borrevik, 2019; Calliari, Surminski, & Mysiak, 2019; Mrema & Ramakrishna, 2010). Further, coalition positions are necessarily compromise positions that need to be negotiated. Since power asymmetries also exist within coalitions, this compromise position may not equally reflect every coalition member's preferences (Jones, Deere-Birkbeck, & Woods, 2010; Narlikar, 2003). Even if the group position carries more weight, it may be rather far from an individual member's national position (Costantini, Sforna, & Zoli, 2016; DeSombre, 2000; Tobin, Schmidt, Tosun, & Burns, 2018).

Under which conditions do the benefits of coalition formation outweigh its costs? When do states create, or join, coalitions? Which coalitions are more successful, and why? Although coalitions are so central to the functioning of multilateral negotiations, they have received surprisingly little academic attention. Coalition formation, maintenance, and effectiveness are not well-understood in multilateral (climate) negotiations (Blaxekjær & Nielsen, 2015; Drahos, 2003; Gray, 2011), partly because negotiation research and scholarship on climate negotiations are disparate fields (Crump & Downie, 2015). As Carter (2015, p. 217) writes, "[d]espite the importance of coalitions in climate change negotiations, there remains a lacuna in the literature on coalition-building and coalition diplomacy in the regime more broadly".

What do we know about coalitions in multilateral negotiations so far? There is some research on the negotiation strategies of small states, for which coalitions are of particular relevance, for example at the United Nations (Albaret & Placidi-Frot, 2016; Panke, 2012, 2013; Thorhallsson, 2012) or within the European Union (Panke, 2011; Thorhallsson, 2016; Thorhallsson & Wivel, 2006). Several studies focus on (developing country) coalitions in world trade negotiations (Cepaluni, Lopes Fernandes, Trecenti, & Damiani, 2014; Costantini et al., 2007; Drahos, 2003; Jones et al., 2010; Lee, 2009; Narlikar, 2003; Narlikar & Odell, 2006; Odell, 2006; Oduwole, 2012; Singh, 2006), or – more rarely – other UN negotiations

(Hampson & Reid, 2003). A number of studies also focus explicitly on environmental negotiations (Allan & Dauvergne, 2013; Mrema & Ramakrishna, 2010; Williams, 2005), including those on climate change.

Research on the climate change negotiations has in particular examined the strategies, challenges, and achievements of individual coalitions, notably the Group of 77 and China (G77) (Chan, 2013; Kasa, Gullberg, & Heggelund, 2008; Vihma, Mulugetta, & Karlsson-Vinkhuyzen, 2011); the Alliance of Small Island States (AOSIS) (Betzold, 2010; Chasek, 2005; de Águeda Corneloup & Mol, 2014; Deitelhoff & Walbott, 2012; Ronneberg, 2016); the emerging economies, Brazil, South Africa, India, and China (BASIC) (Brütsch & Papa, 2013; Downie & Williams, 2018; Hallding, Jürisoo, Carson, & Atteridge, 2013; Hallding et al., 2011; Happaerts, 2015; Hochstetler & Milkoreit, 2014; Hurrell & Sengupta, 2012); and the European Union (Afionis, 2011, 2017; Bäckstrand & Elgström, 2013; Groen & Niemann, 2013). The EU, however, is a special case, since the EU, as a supranational organisation, is itself a party to the Convention. There are also case studies of other coalitions (Atela et al., 2017; Bhandary, 2017; Kameri-Mbote, 2016; Watts & Depledge, 2018). Further, a number of studies use discourse analysis to identify common positions or narratives across countries, to identify potential coalitions, and/or match positions and existing coalitions (Costantini et al., 2016; Jernnäs & Linnér, 2019; Stephenson, Oculi, Bauer, & Carhuayano, 2019; Tobin et al., 2018).

The present volume builds on this body of work, and in particular seeks to widen the focus by examining a plurality of coalitions, and by exploring developments over time. This seems particularly relevant because the climate change negotiations have changed substantially since the first summit in 1995. While coalitions have existed in the climate change negotiations since their inception, this landscape of climate coalitions has fundamentally changed over time, and the number of groups active in climate negotiations has multiplied. Already in 2004, Yamin and Depledge (2004, p. 34) noted that

> The post-Kyoto negotiations have seen a proliferation of new negotiating coalitions, with several groups having emerged over the past few years. This reflects the growing maturity of the regime, accompanied by an increasing awareness among countries of their specific and group interests relative to climate change, along with their desire to participate more actively in the regime. The demand by countries to form new coalitions also responds to the growing tendency of structuring negotiations based on coalitions.

Between 2004 and today, the number of climate coalitions has increased even further, from a handful of negotiating groups at the first COP in 1995 to around 20 at the Paris COP in 2015 (Carter, 2018). Today, there are a wide range of different, partly overlapping coalitions (see Figure 1.1).

How can we make sense of these changing coalition dynamics? Why did these coalitions emerge? Who joins which coalitions, and why? How do coalitions work? How successful are they? And how do they impact the negotiation

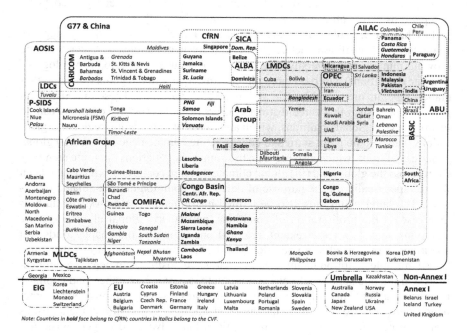

Figure 1.1 Coalitions in the climate change negotiations. Updated from Haller (2018).

process and outcomes? The present volume addresses these questions. It seeks to explore the landscape of climate coalitions and to understand more comprehensively the origins, roles, and effects of coalitions, and coalition proliferation, in the UNFCCC process. In Part I, the book takes a comprehensive approach to coalitions. The contributions in this Part serve as the conceptual and theoretical framework of this book, and do not examine individual coalitions, but rather are interested in general patterns in coalition formation and dynamics. In Part II, the book then turns to individual coalitions. The contributions here try to understand how and why individual coalitions were formed, how they work and with what results. Below, we outline the individual chapters of this volume in more detail.

Outline of contributions

Part I: Coalition formation and behaviour

Part I provides the point of departure and the overall framework for the subsequent chapters by mapping out and taking stock of patterns of cooperation in the climate change negotiations since their inception in 1995. This Part defines coalitions as a specific form of cooperation or diplomacy, different from other negotiation strategies and behaviours. Yet coalitions come in many forms and shapes, and the first part of the edited volume explores this diversity of cooperation patterns across all countries and over time.

In **Chapter 2**, Paula Castro and Carola Klöck seek to systematically map and characterise 25 coalitions active in the climate negotiations to understand the growing number and diversity of coalitions. The authors develop a typology of (climate) coalitions, based on three dimensions: a coalition's geographic and thematic *scope*; its membership *size*; and its level of *formality*. These dimensions in part correlate, such that three distinct clusters or categories of coalitions can be identified: (i) regional coalitions, which are typically pre-existing regional organisations that at some point started to engage in the climate negotiations; (ii) global generic coalitions, which also pre-date the climate negotiations but unite members from across the globe; and (iii) global climate-specific coalitions, which have been formed specifically for the purpose of defending common climate-related objectives. The latter could thus also be termed instrumental coalitions. The authors then discuss the proliferation of coalitions, particularly global climate-specific coalitions, and, to a lesser extent, regional ones. Finally, the chapter finds that coalitions persist, leading to additional and overlapping coalition memberships. These multiple coalition memberships could have positive and negative implications. On the one hand, coalitions can mutually support each other, and common members can forge new alliances and build bridges across coalitions. On the other hand, there are also logistical challenges of multiple coalition memberships, as well as potential tensions between coalitions that advance divergent positions. Which of these effects plays out is an open empirical question.

Chapter 3 directly builds on the typology developed in Chapter 2. Florian Weiler and Paula Castro argue that coalition characteristics make a coalition more or less central to the overall negotiations. In particular, the authors hypothesise that closer coordination within a coalition leads to higher centrality, as does size. In other words, more cohesive and larger coalitions should be more influential. The authors systematically test for differences in coalition behaviour by conceptualising the negotiations as a *network* of negotiation exchanges. Their dataset covers negotiations until 2013, and for this period, the authors find that coalition type does have an impact on coalition behaviour. Notably, regional, climate-focused, larger, and older coalitions tend to play a more central role in the negotiations, in terms of both their levels of activity and popularity, and the way in which they build bridges between their members and all other parties ("betweenness"). In addition, regional, climate-focused, and larger coalitions seem to adopt a position that is closer to more players in the negotiations ("closeness").

Finally, Nicholas Chan takes a more historic and descriptive approach to understanding coalition building and maintenance in **Chapter 4**. His analysis draws on historical institutionalism and focuses on the timing and sequencing of coalition formation of subgroups within the larger Group of 77 and China (G77). Since the UNFCCC process was created by the United Nations General Assembly, the G77 became the "default" coalition for the countries of the Global South. However, as the negotiations developed and were increasingly structured by coalitions, many countries felt the need to create new coalitions. Yet, the G77 dominated the "political space" available for subsequent coalition formation; rather than new, cross-cutting coalitions, this meant that new coalitions were understood to be subgroups of G77, and membership could only be "layered" on top of G77 membership. This means that (developing country) coalitions will seek to associate themselves with the G77 position, but also that they seek to influence this G77 position during G77 coordination meetings. By tracing these dynamics over time, the chapter shows how important historical legacies and institutional context are to understanding current coalition patterns: temporal sequencing has causal significance for the patterns of coalition formation and development that follow. A historical approach to the UNFCCC process thus also provides a more nuanced understanding of the nature of North–South differences in international climate politics.

Part II: Case studies of individual coalitions

Part II of the book zooms in on individual coalitions. In particular, it focuses on some "new" political groups that have emerged in the last rounds of negotiations around the 2009 Copenhagen Accord and the 2015 Paris Agreement. While there are too many coalitions for all of them to be covered, we have selected coalitions that represent the variety of groups active in the climate change regime, including: regional groups and global "meta-coalitions"; formal negotiation groups and informal ad-hoc groups; and coalitions of small and less influential states and cooperation between large and influential ones.

In **Chapter 5**, George Carter traces the involvement of Pacific small island developing states (SIDS) over 30 years of climate change negotiations. Despite being some of the smallest countries in the world, Pacific SIDS have managed to actively engage in and shape the global climate negotiations. The chapter examines the role of these "pivotal players" in three negotiation periods: the early years from 1989 to the signing of the Convention in 1992; an implementation period from the first COP to 2013; and the negotiations leading to the 2015 Paris Agreement and beyond. Pacific SIDS, notably Vanuatu and Tuvalu, co-founded the Alliance of Small Island States (AOSIS) as the core coalition of SIDS and helped to spread awareness of the plight of these "frontline states". Pacific SIDS also influenced AOSIS' negotiating positions and strategies, notably when they held the rotating AOSIS chair. Yet over time, Pacific SIDS also turned to alternative venues to spread their message, and joined or became active in other coalitions. Finally, they increasingly also worked as Pacific SIDS in the negotiations, as a subgroup of AOSIS. In the negotiations around the Paris Agreement, three Pacific island states stand out: Tuvalu opposed the USA on the issue of loss and damage; the Marshall Islands initiated the High Ambition Coalition to pave the way for the Paris Agreement; and Fiji was the first island state to preside over a COP. Clearly, after 30 years of leadership and partnerships, Pacific island states have been, and will continue to be, pivotal players in global climate change negotiations.

Chapter 6 turns to a special type of coalition or group, the Cartagena Dialogue on Progressive Action. Lau Øfjord Blaxekjær examines how this platform for dialogue and open exchange re-created trust in the multilateral UNFCCC process and made it possible for the climate negotiations to resume after the failure of the Copenhagen Summit (COP15). The Cartagena Dialogue is not a "normal" coalition: it does not intervene in the negotiations as a group, and it does not make joint statements or media appearances. Instead, it is a *dialogue*, an informal space for member states from both the Global North and the Global South to come together, exchange viewpoints, learn about others' positions, and share information. As such, the Cartagena Dialogue is best understood as a *community of practice* whose members share an understanding of what is at stake in the UNFCCC process and a common desire for progressive climate policies. The Cartagena Dialogue spans the classical North–South boundary. Members meet regularly face-to-face, but do not agree on consensus positions; rather, members can use notes as they see fit and often refer to, and mutually support, other members' interventions in the negotiations. Through such practices, the Cartagena Dialogue was crucial to taking the negotiations forward and finding compromise at a decisive and difficult moment in the history of the UNFCCC, helping to pave the way for the 2015 Paris Agreement.

Chapter 7 analyses the group of Like-Minded Developing Countries (LMDCs), which emerged in 2012 in the run-up to the 2015 Paris Summit. Lau Øfjord Blaxekjær, Bård Lahn, Tobias Dan Nielsen, Lucia Green-Weiskel, and Fang Fang focus on this period, in which the LMDCs were very vocal. Starting from a constructivist, narrative approach to international relations, they use interviews and other sources to understand how the LMDCs see themselves and how they are seen by others. Their analysis identifies four core characteristics of the LMDCs' narrative position: first, the LMDCs are firmly anchored in the G77 and consider themselves as the "true" representatives of the Global South as a whole. Second, LMDCs see themselves as guardians of the Convention. They insist on maintaining a differentiation between developed Annex I countries, and developing non-Annex I countries, and object to binding emission reductions commitments for all countries. The developed countries' historical responsibility, equity, and the principle of "common but differentiated responsibilities and respective capabilities" (CBDR-RC) – all enshrined in the Convention – imply differential treatment of developed versus developing countries. This relates to the third element of the LMDC position, namely that developing countries are the victims, not the culprits, of anthropogenic climate change. Fourth, and finally, even if LMDCs resist binding emission reduction commitments, they are by no means "blockers", but actively engage in climate action at home. Eventually, the Paris Agreement largely left behind the rigid differentiation of countries into developed and developing countries. Although LMDCs' key demands were not met, the coalition continues to meet and coordinate in the climate change negotiations.

Chapter 8 explores the African Group of Negotiators (AGN), the only UN regional group that is active in substantive negotiations. Simon Chin-Yee, Tobias Dan Nielsen, and Lau Øfjord Blaxekjær argue that the AGN has come a long way from being marginalised in the negotiations, and now plays a significant role. African countries are strongly affected by climate change, but contribute minimally to global greenhouse gas emissions. Based on this vulnerability and lack of (historical) responsibility, the AGN seeks to achieve better representation for Africa's priorities by portraying itself as a coherent and unified coalition – although this does not deny the huge diversity found on the African continent. As the authors show, it has not always been easy to bring together the 54 countries of Africa and speak with one voice in the negotiations. Overlapping coalition memberships, split loyalties, and tensions among African countries exist. Yet, African countries recognise the importance of speaking as one continent; additionally, individual negotiators have become key figures and driven the African agenda in the negotiations. These factors all contributed to the growing role of Africa, through the AGN, in the climate negotiations.

In **Chapter 9**, Joshua Watts examines the role of Latin American and Caribbean countries in the negotiations, by comparing and contrasting the structure, positions, and impact of ALBA, the Bolivarian Alliance of the Peoples of Our America, and AILAC, the Independent Association of Latin American and Caribbean Countries. ALBA and AILAC emerged in 2009 and 2012, respectively, and represent two contrasting voices from the region. ALBA has taken a more controversial approach and sees capitalism and neoliberalism as the root cause of climate change. Accordingly, the coalition has adopted strong and uncompromising positions on equity and climate justice, emphasising the historical responsibility – or historical "debt" – of developed countries for climate change, and objecting market mechanisms. In contrast, AILAC has taken a conciliatory and pragmatic approach (more in line with the Cartagena Dialogue, in which AILAC members participated). AILAC seeks to build bridges, increase trust, and enable compromise and consensus for ambitious and progressive climate policies. The coalition thus encourages mitigation from developing countries alongside emission cuts from developed countries, including through market mechanisms. The overall negotiations have rather leaned toward AILAC's positions. This development, as well as the more formalised structure of AILAC as compared to ALBA, may explain why AILAC is overall more active and engaged in the negotiations, while ALBA's participation and engagement has declined.

The concluding **Chapter 10** returns to the point of departure of this volume, mentioned earlier: the central, but understudied, role of coalitions in multilateral negotiations. Florian Weiler, Paula Castro, and Carola Klöck focus on recurring themes of the volume's diverse contributions. In particular, the authors note four results: first, coalitions are context-specific and need to be studied and understood against the backdrop of overall negotiation dynamics. Coalitions shape negotiations, but negotiations in turn shape coalitions. Second, coalitions tend to be "sticky" and persist. Once created, they tend to remain – even if their level of activity and influence may change over time. Third, coalitions operate at different levels. We note the creation of both sub-groups that are anchored in core coalitions such as the G77 or AOSIS, but also "meta-coalitions" that specifically seek to unite negotiators from across the different coalitions. Fourth, the proliferation of coalitions inevitably leads to multiple and partially overlapping coalition memberships: most countries belong to more than just one coalition. These multiple memberships have both negative and positive effects, creating tensions on the one hand, but also synergies and partnerships on the other. Overall, while the book hopes to contribute to a more comprehensive understanding of coalitions in multilateral (climate) negotiations, many open questions remain, and the concluding chapter also outlines ways forward for coalition research.

Finally, the **Appendix** contains two additional documents: Appendix I lists all countries and the coalitions in which they participate, while Appendix II provides a brief description of all major coalitions that are, or were, active in the climate change coalitions.

References

Afionis, S. (2011). The European Union as a negotiator in the international climate change regime. *International Environmental Agreements: Politics, Law and Economics, 11,* 341–360.

Afionis, S. (2017). *The European Union in International Climate Change Negotiations.* London: Routledge.

Agndal, H. (2007). Current Trends in Business Negotiation Research. An Overview of Articles Published 1996–2005. In *SSE/EFI Working Paper Series in Business Administration, No. 2007:003.* Stockholm: Stockholm School of Economics.

Albaret, M., & Placidi-Frot, D. (2016). Les petits états au Conseil de sécurité: Des strapontins à l'avant-scène. *Critique Internationale, 71,* 19–38.

Allan, I. J., & Dauvergne, P. (2013). The Global South in environmental negotiations: The politics of coalitions in REDD+. *Third World Quarterly, 34*(8), 1307–1322.

Atela, J. O., Quinn, C. H., Arhin, A. A., Duguma, L., & Mbeva, K. L. (2017). Exploring the agency of Africa in climate change negotiations: The case of REDD+. *International Environmental Agreements: Politics, Law and Economics, 17,* 473–482.

Bäckstrand, K., & Elgström, O. (2013). The EU's role in climate change negotiations: From leader to 'leadiator'. *Journal of European Public Policy, 20*(10), 1369–1386.

Bäckstrand, K., Kuyper, J. W., Linnér, B.-O., & Lövbrand, E. (2017). Non-state actors in global climate governance: From Copenhagen to Paris and beyond. *Environmental Politics, 26*(4), 561–579.

Bailer, S. (2004). Bargaining success in the European Union: The impact of exogenous and endogenous power resources. *European Union Politics, 5*(1), 99–123.

Bandyopadhyay, S., & Chatterjee, K. (2012). Models of Coalition Formation in Multilateral Negotiations. In R. Croson & G. E. Bolton (Eds.), *The Oxford Handbook of Economic Conflict Resolution* (pp. 77–89). Oxford: Oxford University Press.

Bazermann, M. H., Curhan, J. R., Moore, D. A., & Valley, K. L. (2000). Negotiation. *Annual Review of Psychology, 51,* 279–314.

Betzold, C. (2010). 'Borrowing' power to influence international negotiations: AOSIS in the climate change regime, 1990–1997. *Politics, 30*(3), 131–148.

Bhandary, R. R. (2017). Coalition strategies in the climate negotiations: An analysis of mountain-related coalitions. *International Environmental Agreements: Politics, Law and Economics, 17*(2), 173–190.

Blaxekjær, L. Ø., & Nielsen, T. D. (2015). Mapping the narrative positions of new political groups under the UNFCCC. *Climate Policy, 15*(6), 751–766.

Borrevik, C. A. (2019). *"We started climate change": A multi-level ethnography of pacific climate leadership* (PhD thesis). University of Bergen, Bergen.

Brütsch, C., & Papa, M. (2013). Deconstructing the BRICS: Bargaining coalition, imagined community, or geopolitical fad? *The Chinese Journal of International Politics, 6,* 299–327.

Buchner, B., & Carraro, C. (2006). Parallel Climate Blocs: Incentives to Cooperation in International Climate Negotiations. In Working Paper No. 45/WP/2006. Venice: Department of Economics, Ca'Foscari University of Venice.

Calliari, E., Surminski, S., & Mysiak, J. (2019). The Politics of (and Behind) the UNFCCC's Loss and Damage Mechanism. In R. Mechler, L. Bouwer, T. Schinko, S. Surminski, & J. Linnerooth-Bayer (Eds.), *Loss and Damage from Climate Change: Climate Risk Management, Policy and Governance* (pp. 155–178). Cham: Springer.

Carter, G. (2015). Establishing a Pacific Voice in the Climate Change Negotiations. In G. Fry & S. Tarte (Eds.), *The New Pacific Diplomacy* (pp. 205–220). Canberra: ANU Press.

Carter, G. (2018). *Multilateral consensus decision making: How Pacific island states build and reach consensus in climate change negotiations* (Doctor of Philosophy). Australian National University, Canberra.

Cepaluni, G., Lopes Fernandes, I. F. d. A., Trecenti, J. A. Z., & Damiani, A. P. (2014). "United We Stand, Divided We Fall": Which countries join coalitions more often in GATT/WTO negotiations?

Chan, N. (2013). *The construction of the South: Developing countries, coalition formation and the UN climate change negotiations, 1988–2012* (DPhil). University of Oxford, Oxford.

Chasek, P. (2005). Margins of power: Coalition building and coalition maintenance of the South Pacific island states and the alliance of small island states. *Review of European Community and International Environmental Law, 14*(2), 125–137.

Ciplet, D., Khan, M., & Roberts, J. T. (2015). *Power in a Warming World: The New Global Politics of Climate Change and the Remaking of Environmental Inequality.* Boston: MIT Press.

Costantini, V., Crescenzi, R., Filippis, F. D., & Salvatici, L. (2007). Bargaining coalitions in the WTO agricultural negotiations. *The World Economy, 30*(5), 863–891.

Costantini, V., Sforna, G., & Zoli, M. (2016). Interpreting bargaining strategies of developing countries in climate negotiations. A quantitative approach. *Ecological Economics, 121,* 128–139.

Crump, L., & Downie, C. (2015). Understanding climate change negotiations: Contributions from international negotiation and conflict management. *International Negotiation, 20,* 146–174.

Crump, L., & Zartman, I. W. (2003). Multilateral negotiation and the management of complexity. *International Negotiation, 8,* 1–5.

de Águeda Corneloup, I., & Mol, A. P. J. (2014). Small island developing states and international climate change negotiations: The power of moral "leadership". *International Environmental Agreements: Politics, Law and Economics, 14*(3), 281–297.

Deitelhoff, N., & Walbott, L. (2012). Beyond soft balancing: Small states and coalition-building in the ICC and climate negotiations. *Cambridge Review of International Affairs, 25*(3), 345–366.

DeSombre, E. R. (2000). Developing country influence in global environmental negotiations. *Environmental Politics, 9*(3), 23–42.

Downie, C., & Williams, M. (2018). After the Paris agreement: What role for the BRICS in global climate governance? *Global Policy, 9*(3), 398–407.

Drahos, P. (2003). When the weak bargain with the strong: Negotiations in the World Trade Organization. *International Negotiation, 8,* 79–109.

Dupont, C. (1994). Coalition Theory: Using Power to Build Cooperation. In I. W. Zartman (Ed.), *International Multilateral Negotiations: Approaches to the Management of Complexity* (pp. 148–177). San Francisco: Jossey-Bass Publishers.

Dupont, C. (1996). Negotiation as coalition building. *International Negotiation, 1*(1), 47–64.

Elgström, O., Bjurulf, B., Johansson, J., & Sannerstedt, A. (2001). Coalitions in European Union negotiations. *Scandinavian Political Studies, 24*(2), 111–128.

Finke, D. (2012). Proposal stage coalition-building in the European Parliament. *European Union Politics, 13*(4), 487–512.

Gamson, W. A. (1961). A theory of coalition formation. *American Sociological Review, 26*(3), 373–382.

Gaventa, J. (2010). Environmental mega-conferences and climate governance beyond the nation state: A Bali case study. *St. Anthony's International Review, 5*(2), 29–45.

Gray, B. (2011). The complexity of multiparty negotiations: Wading into the muck. *Negotiation and Conflict Management Research, 4*(3), 169–177.

Groen, L.', & Niemann, A. (2013). The European Union at the Copenhagen climate negotiations: A case of contested EU actorness and effectiveness. *International Relations, 27*(3), 308–324.

Hallding, K., Jürisoo, M., Carson, M., & Atteridge, A. (2013). Rising powers: The evolving role of BASIC countries. *Climate Policy, 13*(5), 608–631.

Hallding, K., Olsson, M., Atteridge, A., Vihma, A., Carson, M., & Román, M. (2011). Together Alone: BASIC Countries and the Climate Change Conundrum. In *TemaNord 2011:530.* Copenhagen: Nordic Council of Ministers.

Haller, J. (2018). Party Groupings in the UNFCCC. In *Wikimedia Commons.*

Hampson, F. O., & Reid, H. (2003). Coalition diversity and normative legitimacy in human security negotiations. *International Negotiation, 8,* 7–42.

Happaerts, S. (2015). Rising Powers in Global Climate Governance: Negotiating Inside and Outside the UNFCCC. In D. Lesage & T. Van de Graaf (Eds.), *Rising Powers and Multilateral Institutions* (pp. 238–257). Basingstoke: Palgrave Macmillan.

Hochstetler, K., & Milkoreit, M. (2014). Emerging powers in the climate negotiations: Shifting identity conceptions. *Political Research Quarterly, 67*(1), 224–235.

Hurrell, A., & Sengupta, S. (2012). Emerging powers, North–South relations and global climate politics. *International Affairs, 88*(3), 463–484.

Jernnäs, M., & Linnér, B.-O. (2019). A discursive cartography of nationally determined contributions to the Paris climate agreement. *Global Environmental Change, 55,* 73–83.

Jones, E., Deere-Birkbeck, C., & Woods, N. (2010). *Manoeuvring at the Margins: Constraints Faced by Small States in International Trade Negotiations.* London: Commonwealth Secretariat.

Kameri-Mbote, P. (2016). The Least Developed Countries and Climate Change Law. In K. R. Gray, R. Tarasofsky, & C. Carlarne (Eds.), *The Oxford Handbook of International Climate Change Law* (pp. 740–760). Oxford: Oxford University Press.

Kasa, S., Gullberg, A. T., & Heggelund, G. (2008). The group of 77 in the international climate negotiations: Recent developments and future directions. *International Environmental Agreements: Politics, Law and Economics, 8*(2), 113–127.

Lee, D. (2009). Bringing an Elephant into the Room: Small African State Diplomacy in the WTO. In A. F. Cooper & T. M. Shaw (Eds.), *The Diplomacies of Small States: Between Vulnerability and Resilience* (pp. 195–206). Basingstoke: Palgrave Macmillan.

Lövbrand, E., Hjerpe, M., & Linnér, B.-O. (2017). Making climate governance global: How UN climate summitry comes to matter in a complex climate regime. *Environmental Politics, 26*(4), 580–599.

Mrema, E., & Ramakrishna, K. (2010). The Importance of Alliances, Groups and Partnerships in International Environmental Negotiations. In T. Honkonen & E. Couzens (Eds.), *International Environmental Law-Making and Diplomacy Review 2009* (pp. 183–192). Joensuu: University of Eastern Finland.

Narlikar, A. (2003). *International Trade and Developing Countries: Bargaining Coalitions in the GATT & WTO.* London and New York: Routledge.

Narlikar, A., & Odell, J. S. (2006). The Strict Distributive Strategy for a Bargaining Coalition: The Like Minded Group in the World Trade Organization. In J. S. Odell (Ed.), *Negotiating Trade: Developing Countries in the WTO and NAFTA* (pp. 115–144). Cambridge: Cambridge University Press.

Odell, J. S. (2006). *Negotiating Trade: Developing Countries in the WTO and NAFTA.* Cambridge: Cambridge University Press.

Oduwole, J. (2012). An Appraisal of Developing Country Coalition Strategy in the WTO Doha Round Agriculture Negotiations. In Society of International Economic Law Working Paper No. 2012/54. Singapore.

Panke, D. (2011). Small states in EU negotiations: Political dwarfs or power-brokers? *Cooperation and Conflict, 46*(2), 123–143.

Panke, D. (2012). Dwarfs in international negotiations: How small states make their voices heard. *Cambridge Review of International Affairs, 25*(3), 313–328.

Panke, D. (2013). *Unequal Actors in Equalising Institutions: Negotiations in the United Nations General Assembly.* Basingstoke: Palgrave Macmillan.

Penetrante, A. M. (2013). Common but Differentiated Responsibilities: The North–South Divide in the Climate Change Negotiations. In G. Sjöstedt & A. M. Penetrante (Eds.), *Climate Change Negotiations: A Guide to Resolving Disputes and Facilitating Multilateral Cooperation* (pp. 249–276). Abingdon and New York: Routledge.

Ronneberg, E. (2016). Small Islands and the Big Issue: Climate Change and the Role of the Alliance of Small Island States. In K. R. Gray, R. Tarasofsky, & C. Carlarne (Eds.), *The Oxford Handbook of International Climate Change Law* (pp. 761–778). Oxford: Oxford University Press.

Rubin, J. Z., & Zartman, I. W. (2000). *Power and Negotiation.* Ann Arbor: University of Michigan Press.

Sagi, E., & Diermeier, D. (2017). Language use and coalition formation in multiparty negotiations. *Cognitive Science, 41*, 259–271.

Schmidt, A., Ivanova, A., & Schäfer, M. S. (2013). Media attention for climate change around the world: A comparative analysis of newspaper coverage in 27 countries. *Global Environmental Change, 23*(5), 1233–1248.

Schroeder, H., & Lovell, H. (2012). The role of non-nation state actors and side events in the international climate negotiations. *Climate Policy, 12*(1), 23–37.

Singh, J. P. (2006). Coalitions, developing countries, and international trade: Research findings and prospects. *International Negotiation, 11*, 499–514.

Starkey, B., Boyer, M. A., & Wilkenfeld, J. (2005). *Negotiating a Complex World: An Introduction to International Negotiation* (2nd ed.). Lanham, MD: Rowman & Littlefield.

Stephenson, S. R., Oculi, N., Bauer, A., & Carhuayano, S. (2019). Convergence and divergence of UNFCCC nationally determined contributions. *Annals of the American Association of Geographers, 109*(4), 1240–1261.

Stevenson, W. R. (1985). The concept of "coalition" in organization theory and research. *Academy of Management Review, 10*(2), 256–268.

Thorhallsson, B. (2012). Small states in the UN security council: Means of influence? *The Hague Journal of Diplomacy, 7*, 135–160.

Thorhallsson, B. (2016). *The Role of Small States in the European Union* (2nd ed.). Abingdon: Routledge.

Thorhallsson, B., & Wivel, A. (2006). Small states in the European Union: What do we know and what would we like to know? *Cambridge Review of International Affairs, 19*(4), 651–668.

Tobin, P., Schmidt, N. M., Tosun, J., & Burns, C. (2018). Mapping states' Paris climate pledges: Analysing targets and groups at COP21. *Global Environmental Change, 48*, 11–21.

UNFCCC. (1992). United Nations Framework Convention on Climate Change. In *Contained in Document FCCC/INFORMAL/84.*

United Nations. (1988). Protection of Global Climate for Present and Future Generations of Mankind. In *United Nations General Assembly Resolution A/RES/43/53*. New York: United Nations.

Vihma, A., Mulugetta, Y., & Karlsson-Vinkhuyzen, S. (2011). Negotiating solidarity? The G77 through the prism of climate change negotiations. *Global Change, Peace & Security, 23*(3), 315–334.

Watts, J., & Depledge, J. (2018). Latin America in the climate change negotiations: Exploring the AILAC and ALBA coalitions. *Wiley Interdisciplinary Reviews: Climate Change, 9*(6), e533.

Williams, M. (2005). The third world and global environmental negotiations: Interests, institutions and ideas. *Global Environmental Politics, 5*(3), 48–69.

Wu, J., & Thill, J.-C. (2018). Climate change coalition formation and equilibrium strategies in mitigation games in the post-Kyoto era. *International Environmental Agreements: Politics, Law and Economics, 18*(4), 573–598.

Yamin, F., & Depledge, J. (2004). *The International Climate Change Regime: A Guide to Rules, Institutions and Procedures*. Cambridge: Cambridge University Press.

Part I

Overview

Coalition dynamics in the climate change negotiations

2 Fragmentation in the climate change negotiations

Taking stock of the evolving coalition dynamics

Paula Castro and Carola Klöck

Introduction

Multilateral negotiations, such as those under the United Nations Framework Convention on Climate Change (UNFCCC), are structured by coalitions. As soon as more than two parties negotiate, parties join forces and cooperate with others in order to reach common goals (Dupont, 1996). Although such coalitions have been a key feature of the climate change negotiations since their inception in 1990, coalition dynamics have changed significantly over almost 30 years of the climate change negotiations. Particularly since the 1997 Kyoto Protocol entered into force in 2005 and discussions started to revolve around a successor agreement, a plethora of new coalitions has emerged, with partly overlapping memberships and positions. How can we make sense of this growing fragmentation of the climate change negotiations? How do these coalitions differ from, or how are they similar to, each other?

Although coalitions are central to understanding negotiation dynamics, coalitions in the climate change negotiations have received relatively scant academic attention. Research to date mainly consists of case studies of individual coalitions, notably the European Union (Afionis, 2011; Groen & Niemann, 2013), the Group of 77 and China (G77) (Kasa, Gullberg, & Heggelund, 2008; Vihma, Mulugetta, & Karlsson-Vinkhuyzen, 2011), the Alliance of Small Island States (AOSIS) (Benjamin, 2011; Betzold, 2010; Betzold, Castro, & Weiler, 2012; de Águeda Corneloup & Mol, 2014; Ronneberg, 2016), Brazil-South Africa-India-China (BASIC) (Downie & Williams, 2018; Hallding, Jürisoo, Carson, & Atteridge, 2013; Hallding et al., 2011) and other coalitions (Atela, Quinn, Arhin, Duguma, & Mbeva, 2017; Bhandary, 2017; Blaxekjær & Nielsen, 2015; Downie & Williams, 2018; Edwards & Roberts, 2015; Kameri-Mbote, 2016; Roger, 2013; Watts & Depledge, 2018). The contributions of this volume add to these case studies of coalitions. A second strand of research compares the positions of individual countries – typically based on their nationally determined contributions (NDCs) or on written position papers submitted to the UNFCCC – to assess the cohesion of existing coalitions, or to identify shared narratives or "storylines" and "discourse coalitions" (Blaxekjær & Nielsen, 2015; Costantini, Sforna, & Zoli, 2016; Jernnäs & Linnér, 2019; Stephenson, Oculi, Bauer, & Carhuayano, 2019;

Tobin, Schmidt, Tosun, & Burns, 2018; Woods & Kristófersson, 2016). Overall, however, coalition dynamics in multilateral (climate) negotiations have not yet been addressed comprehensively (Blaxekjær & Nielsen, 2015; Laatikainen, 2017; Laatikainen & Smith, 2020b; Narlikar, 2003). Carter (2015, p. 217) thus notes "a lacuna in the literature on coalition-building and coalition diplomacy in the regime more broadly". We still understand coalition formation, maintenance, and change poorly, in climate negotiations and beyond.

In order to address this gap, this is, in order to understand the changing coalition dynamics and growing fragmentation of the UNFCCC process, as well the implications of this fragmentation, we need to first systematically map and characterise coalitions. This chapter therefore seeks to first describe and understand the diversity of coalitions that have been involved in the climate change negotiations over time (thereby also establishing some of the concepts and ideas that underpin the following chapters on individual coalitions), and then to use this description to help us make sense of the observed evolution in the landscape of UNFCCC coalitions over time. In the second section, we hence draw on negotiation and coalition theory to identify three dimensions – scope (geographic and thematic), membership size, and level of formality – that serve to describe coalitions. Based on these criteria, we then identify three clusters of coalitions: regional; global generic; and global climate-specific coalitions. In the third section, we then discuss to what extent this typology of coalitions helps us to understand the growing fragmentation of the UNFCCC process, as well as its implications for the overall negotiation process and outcome. In the fourth section, we conclude.

Characterising climate coalitions

Multilateral negotiations are complex, being "multi-parties, multi-issues, multi-roles, and multi-values" (Muldoon, 2005, p. 11). Coalitions reduce this complexity by reducing the number of actors. At the same time, coalitions increase the bargaining power of their members since a position backed by many parties carries more weight than that of an individual party (Chasek, 2015; Dupont, 1996; Laatikainen & Smith, 2020a). Coalitions, in the context of multilateral negotiations, are groups of countries that explicitly coordinate their positions and pool their resources in order to achieve common goals (Chan, 2013; Dupont, 1996; Odell, 2006; Williams, 2005). Shared interests are thus seen as a precondition for, as well as the purpose of, coalition formation, although the joint coalition position necessarily is a compromise that is different from a member's individual position (Costantini, Crescenzi, Filippis, & Salvatici, 2007; Laatikainen & Smith, 2020b).

There is a wide range of terms to designate coalitions in research and practice. In the climate change negotiations, the UNFCCC lists diverse "party groupings", including, among others, the "African *Group* of Negotiators" (AGN), the "*Alliance* of Small Island States" (AOSIS) or the "*Coalition* for Rainforest Nations" (CfRN) (our emphasis).[1] Research on coalitions similarly speaks about coalitions, negotiating groups, blocs, alliances, communities, or combinations thereof (e.g. Blaxekjær & Nielsen, 2015; Chasek, 2015; Hirsch, 2016). Audet

(2013), for instance, uses coalition as the more generic term and distinguishes between "alliance coalitions" – typically smaller, issue-specific groups with strategic negotiation goals – and "bloc coalitions" – larger, more durable groups based on common identities and ideologies that frequently negotiate across a variety of issues (see also Narlikar, 2003). For Starkey, Boyer, and Wilkenfeld (2005, p. 58), alliances are more formal, permanent, and long-term, while coalitions are temporary and only last for the duration of a negotiation. Ott, Bauer, Brandi, Mersmann, and Weischer (2016, p. 5) use "alliances" as the generic term, defining them as "any grouping that comprises more than two and less than the full multilateral set of countries party to the UNFCCC and that has not reached the degree of institutionalization of an international organization". Within alliances, the authors differentiate between coalitions and blocs. While formal negotiating blocs are more permanent, such as the G77 or AOSIS, coalitions are short-term and informal cooperative efforts of countries from different blocs that join forces to achieve a specific outcome at a specific climate conference, such as the "Green Group" that formed at the very first climate conference in Berlin in 1995 to push for a more ambitious agreement (which was agreed two years later as the 1997 Kyoto Protocol) (Afionis, 2017; Yamin & Depledge, 2004, p. 35). Brütsch and Papa (2013) instead propose to distinguish between tactical coalitions and identity-based communities. Other scholars use adjectives to differentiate coalitions. Bhandary (2017) thus speaks of political coalitions, issue-specific coalitions, and structural or tactical coalitions, without, however, clearly defining these coalition types. For the World Trade Organization (WTO), Drahos (2003) identifies three types of coalitions: regional groups that exist outside of the WTO, sectoral groups that have been formed in response to specific issues, and groups based on UN categories, notably least developed countries (LDCs). Wagner (1999) uses a coalition's level of cohesion (low or high) and what the author refers to as "coalition arrangement" – the extent to which coalition preferences are distinct and clear, or vary – to develop a 2×2 matrix of coalition types.

Clearly, coalitions come in various forms and shapes, yet there is no widely used and clearly defined nomenclature of coalitions in multilateral negotiations. To avoid adding more confusion to an already muddled field, we purposefully abstain from offering new names to different types of coalitions. Rather, we keep the term "coalition" as a broad concept encompassing all of them, and attempt to use very specific and clearly observable characteristics of coalitions in order to categorise them.

The research reviewed above has highlighted a range of dimensions on which coalitions differ, including their *goals*, *scope or reach*, *size*, *level of formality*, *permanence over time*, and *level of cohesion*. Frequently, existing research seems to assume that several of these dimensions tend to co-occur. It tends to distinguish formal, large, durable, and identity-based coalitions, from more temporary, smaller, issue-specific, and strategic ones. There does not seem to be any systematic attempt to assess how frequently these characteristics actually happen in such a constellation.

We therefore first classify coalitions along several of the dimensions listed above independently, and then attempt to build clusters of similar coalitions based on how often these dimensions co-exist.

We consider that the *goals* of a coalition cannot be readily observed from the outside, and that they can change substantially over time. For example, AOSIS probably started as a strategic coalition with a clear aim of highlighting the challenges imposed by climate change on small island states. While this is still its central aim, over time AOSIS has evolved into an identity-based coalition whose members engage with each other beyond strategic objectives, and which is active on a broad variety of issue-areas. For this reason, we abstain from focusing on this dimension in our analysis. We further consider the last two dimensions – *permanence* and *cohesiveness* – not as intrinsic characteristics of coalitions, but rather as phenomena *resulting from* their type of membership, thematic scope, and success in achieving their goals. Therefore, we consider the temporal persistence of coalitions in our discussion section, while their cohesiveness is one of the aspects addressed in Chapter 3 of this volume (Weiler & Castro, 2020). In the following, we describe each of the remaining dimensions (*scope or reach, membership size*, and *level of formality*), and discuss how climate coalitions fare on them.

Scope or reach

A first dimension on which to differentiate coalitions is their "scope" or "reach". We propose two sub-dimensions: geographic scope on the one hand, and thematic scope on the other.

Under the rubric geographic scope, which denotes the geographic coverage of the members of the coalition, we differentiate between regional and global coalitions. The UN system works through regional groups. While these regions remain important for organisational purposes in the climate negotiations, for example, for the location of the yearly climate conferences, they are less important for substantive negotiations, partly because climate interests do not converge along regional lines (Richards, 2001; Yamin & Depledge, 2004). Only Africa corresponds to a UN regional group that acts as a coalition in substantive negotiations, as the African Group of Negotiators (AGN). Nevertheless, many coalitions are based in regions or sub-regions, and are open only to members of that region or sub-region. For example, while there is no group of Latin American and Caribbean countries, both the Bolivarian Alliance for the Peoples of our America (ALBA) and the Independent Association of Latin America and the Caribbean (AILAC) are open to countries of Latin America and the Caribbean only.

Other coalitions include countries from more than one UN region. AOSIS for example unites small island states from the Pacific, Caribbean, Atlantic, and Indian Ocean, as well as Singapore and Timor-Leste. The Coalition for Rainforest Nations (CfRN) has members from Africa, Asia, the Americas, and the Pacific. The G77 represents most of the Global South. Such coalitions that span more than one world region can be considered global in reach.

Beyond geographic scope, coalitions also differ in their thematic scope, or the range of issues that a coalition covers. Here, we differentiate coalitions that are active on climate change only, and coalitions that are active across issues (cf. Cepaluni, Galdino, & de Oliveira, 2012). The origins of such broad, cross-issue, or generic coalitions are not the UNFCCC process, but other international forums or negotiations, yet they have started to also act as a coalition in the climate context. The G77, for example, has its roots in trade negotiations, but is now also active in the UNFCCC. The Central American Integration System (SICA), the Organization of Petroleum Exporting Countries (OPEC), or the European Union (EU) are organisations that pre-date the climate process and coordinate their actions on a diverse range of issues, including (but not limited to) climate change. In a way, these broad coalitions are similar to the "bloc coalitions" in Audet (2013) and Narlikar (2003). However, our classification is more specific, as it focuses only on the thematic scope covered by the group, rather than also covering the motives leading to coalition formation, their level of formality, or their permanence.

In contrast, climate-specific coalitions have been created specifically in the context of the climate change negotiations, and their activities are restricted to these negotiations, at least at the outset. AOSIS was founded to give a voice to island interests in the climate change negotiations. AILAC, the Cartagena Dialogue or the Like-Minded Developing Countries (LMDCs) are other examples of coalitions that were created specifically to advance common positions in the UNFCCC process. Some of these coalitions even focus on individual agenda items within the climate negotiations; the CfRN, for instance, works mostly on forests, or REDD (reducing emissions from deforestation and forest degradation). These "item-specific" coalitions are the exception rather than the rule, however, as coalitions typically coordinate on more than one agenda item, although most coalitions will prioritise specific agenda items. Adaptation, for example, is of particular concern to vulnerable developing countries represented in AOSIS, the Climate Vulnerable Forum (CVF), or LDCs (Carter, 2018).[2]

Membership

A second dimension on which to differentiate coalitions is membership size (Cepaluni et al., 2012). We can roughly distinguish between small and large coalitions. Size is related to cohesion or unity, both factors that are often related to coalition success. On the one hand, there is strength in numbers in multilateral negotiations; larger coalitions are more influential, because they: can more credibly threaten to block consensus; can put pressure on laggards; and have more, and more diverse, resources and capabilities (Cepaluni et al., 2012; Hampson & Reid, 2003; Panke, 2013). On the other hand, the larger a coalition, the more diverse, on average, its membership, and hence the more difficult it is to find and maintain a common position, which may decrease coalition influence (Chasek & Rajamani, 2003; Ciplet, Khan, & Roberts, 2015; Drahos, 2003).

Of course, size is a relative concept, and it is not *a priori* clear which coalitions are small and which are large. In the climate context, coalition size ranges from just four members (e.g. BASIC and the Group of Mountainous Landlocked Developing Countries, MLDCs) to over 100 members (G77 has 134 members, the High Ambition Coalition (HAC) has 107 members). On average, each coalition comprises 29 members, but the median size is 14. We consider coalitions whose membership size is over the median as large, and all others as small. We thus count 14 small and 11 large coalitions.[3]

Level of formality

The third and final feature we use to describe coalitions is their level of formality, or the way in which the coalition is structured and operates. Coalitions can be highly formalised and institutionalised, or can function very loosely and informally.

Some coalition research assumes that coalitions are formed for specific objectives only, and disband when this objective has been achieved. For the WTO, for example, Drahos (2003, p. 88) writes that "group life is characterized by looseness, temporariness and pragmatism". From this viewpoint, there is no formal procedure for entry or exit, or even for creation. There are no founding documents, no international agreements, no secretariat or organisational structure. In this respect, they resemble "coalitions" in the sense of Starkey et al. (2005) and Ott et al. (2016). In contrast, some coalitions are, or resemble, international organisations that act as a coalition in climate (or other) negotiations. These formal coalitions have a founding agreement, and members have to follow a formal procedure to join or leave the coalition, such that the coalition membership is well-defined. There may be a secretariat or other bodies that are tasked with centrally coordinating member positions and outlining the coalition's joint negotiation position. Several coalitions in addition have their own website, where they publish their positions, list their membership, or describe their history. Such coalitions are thus highly institutionalised, often long-lasting, and resemble what Starkey et al. (2005) call "alliances" and Ott et al. (2016) refer to as "blocs".

In the climate change negotiations, several international organisations have started to operate as a coalition, as mentioned earlier, including the EU (which is a special case, given that it itself is a party to the UNFCCC), OPEC, SICA, or the Caribbean Community (CARICOM). But even coalitions that have emerged within the climate context can be formal. AILAC, for example, has been specifically created as a climate coalition, but has a founding document, is recognised as a formal negotiating group under the UNFCCC, and has a clearly defined membership. AOSIS is a similar case of a formal international group arising from the climate change negotiations, which has survived for over 20 years and now engages in other international issue-areas, too.

In contrast, other coalitions are much less institutionalised, coordinate their positions only loosely, and/or have a fluid and changing membership. Sometimes, these coalitions form in an ad-hoc manner and operate only during a short period; other coalitions operate informally, but persist nevertheless. For example, the Cartagena Dialogue was purposefully created as an informal platform for open exchange – hence the name Cartagena *Dialogue* (Blaxekjær & Nielsen, 2015). The Umbrella Group has existed since the start of the negotiations, but operates informally and its membership (and name) has changed over time. The LMDCs or HAC are more recent examples of informal coalitions.

For our classification, we rely on four indicators of formality: (i) whether the coalition has its own website; (ii) whether it defines itself as a formal organization (for example, on its website), or is a recognised regional group, or has official founding documents as an international organisation; (iii) whether it is recognised on the UNFCCC website[4] as a formal negotiation group; and (iv) whether it has a fixed membership. Coalitions that fulfil at least three of these indicators are classified as formal.

Summary: patterns of coalition types

Clearly, coalitions come in many forms and shapes. The three dimensions we propose are not mutually exclusive, and diverse combinations are possible.

Table 2.1 categorises 25 different coalitions that have been active in the UNFCCC process according to negotiation summaries by the *Earth Negotiations Bulletin* (International Institute for Sustainable Development, 1997–2018) on these dimensions. The Table showcases the diversity of coalitions on most of our dimensions, but does not reveal any clear patterns at first sight. To further explore to what extent certain general coalition types exist, Table 2.2 correlates the different dimensions with each other. Not all dimensions are related, but some patterns can be discerned. Regional coalitions are more likely to be generic in their thematic scope. By definition, they are geographically restrictive in terms of their membership. They also tend to be small and formal. Taken together, this suggests that most of these regional coalitions consist of pre-existing regional organizations that at some point started to engage in the climate negotiations as well. AILAC is the exception here (along with CACAM and CG-11, which are, however, no longer active); AILAC resembles these regional organisations, but emerged specifically in the context of the climate negotiations. Since AILAC may well become a regional organisation and work beyond the climate context, we subsume AILAC under the category of regional coalitions and make no further distinctions.

In contrast, global coalitions are typically climate-specific and frequently work informally. There are three exceptions, i.e., global coalitions that are not climate-specific and rather formal: G77, LDCs, and OPEC. Arguably, these three coalitions hence form their own category. Overall, we thus have three categories or clusters of coalitions: (*i*) regional coalitions; (*ii*) global climate-specific coalitions; and (*iii*) global generic coalitions. We could also term our second category (global climate-specific coalitions) more generally as instrumental coalitions, as they have clear climate-related objectives at their core. Our typology thus resembles the distinction Drahos (2003) makes for trade negotiations.

Table 2.1 Categorisation of coalitions in the UNFCCC process according to our dimensions. Bold caps indicate "new" coalitions that emerged after 2005

Cluster	Coalition	Scope		Size	Formality
		Geographic	Thematic		
i	AGN	regional	generic	large	formal
	ALBA	regional	generic	small	formal
	Arab Group	regional	generic	large	formal
	CARICOM	regional	generic	small	formal
	COMIFAC	regional	generic	small	formal
	Congo Basin	regional	generic	small	informal
	EU	regional	generic	large	formal
	SICA	regional	generic	small	formal
	Visegrad	regional	generic	small	formal
	AILAC	regional	climate-specific	small	formal
	CACAM	regional	climate-specific	small	informal
	CG-11	regional	climate-specific	small	informal
ii	AOSIS	global	climate-specific	large	formal
	BASIC	global	climate-specific	small	informal
	Cartagena D.	global	climate-specific	large	informal
	CfRN	global	climate-specific	large	formal
	CVF	global	climate-specific	large	formal
	EIG	global	climate-specific	small	informal
	HAC	global	climate-specific	large	informal
	LMDCs	global	climate-specific	large	informal
	MLDCs	global	climate-specific	small	informal
	Umbrella	global	climate-specific	small	informal
iii	G77 and China	global	generic	large	formal
	LDCs	global	generic	large	formal
	OPEC	global	generic	small	formal

Table 2.2 Correlations between the dimensions. Correlation factors ≥0.5 are highlighted

		Geographic scope (regional)	Thematic scope (generic)	Size (small)	Formality (formal)
Geographic scope	(regional)	1			
Thematic scope	(generic)	**0.52**	1		
Size	(small)	0.37	0.05	1	
Formality	(formal)	0.29	**0.62**	−0.23	1

Discussion

We started this chapter by observing the growing fragmentation of the climate negotiations into ever more coalitions. To what extent does the previous characterisation of these diverse coalitions and the resulting three clusters help us make sense of these changes over time? And how does this fragmentation affect the negotiation process and outcomes?

Proliferation of coalitions

Figure 2.1 tabulates all coalitions over time. For each coalition, we register whether it intervened at a specific Conference of the Parties (COP) according to the *Earth Negotiations Bulletin*. The increase in the number of coalitions operating at the same time is clearly visible. Much of this increase happened after the Kyoto Protocol entered into force in 2005. In particular, the negotiations toward a successor agreement of the Kyoto Protocol at the 2009 Copenhagen COP and the 2015 Paris COP led to the creation of new coalitions, or the activation of regional organisations within the UNFCCC process (see also Groen, 2020), although Yamin and Depledge (2004, p. 34) already note a proliferation of coalitions in the post-Kyoto negotiations.

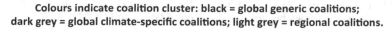

Colours indicate coalition cluster: black = global generic coalitions; dark grey = global climate-specific coalitions; light grey = regional coalitions.

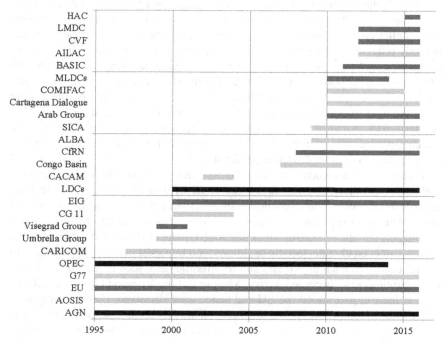

Figure 2.1 Climate coalitions over time.

In terms of the three clusters of coalitions identified in the previous section, these new, post-2005 coalitions belong to two clusters: regional coalitions and global climate-specific, or instrumental, coalitions. In contrast, the three global generic coalitions (OPEC, G77, and LDCs) have all been active in the climate negotiations for a long time.

How should we explain the proliferation of coalitions in the climate negotiations? While each coalition has its own history and raison d'être, factors at the level of the overall negotiations on the one hand, and within coalitions on the other, can explain the creation of new coalitions. First, the overall negotiation dynamics encourage coalition formation (see Chan, 2020). As the UNFCCC process matured, the negotiations increasingly became structured around coalitions, with, for example, smaller consultations such as "Friends of the Chair" groups bringing together representatives from the different coalitions (Yamin & Depledge, 2004). Similarly, when the negotiations stall, countries may seek to forge new alliances to achieve compromises; the Cartagena Dialogue or HAC are thus direct responses to negotiation dynamics, namely the failure of the 2009 Copenhagen COP, and tensions at the 2015 Paris COP, respectively. In addition, the emergence of new issue areas within the climate change negotiations has led to new alliances. This is particularly notable for the case of REDD; once this forest-related issue area emerged, not only was a new global coalition of countries with tropical forests created (CfRN), but also existing regional organisations with a focus on forest issues became active (Congo Basin and COMIFAC).

Second, dynamics within coalitions encourage the creation of sub-coalitions (see Chan, 2020). Naturally, some coalition members are more powerful than others. Smaller countries may feel marginalised, and their interests may not be sufficiently reflected in the group position. To ensure that their voice is heard, these countries may hence decide to form their own (sub-)coalition. For example, the LDCs found their position being sidelined in the G77, and hence started to mobilise as LDCs in 2000 (Yamin & Depledge, 2004, p. 36). Similarly, the Pacific islands sought to establish a stronger voice within AOSIS around the Paris COP (Carter, 2015, 2018). Along the same lines, disagreement within the Umbrella Group's predecessor, JUSSCANNZ, led Switzerland to form the Environmental Integrity Group (Yamin & Depledge, 2004, p. 47f).

Figure 2.1 also clearly indicates the persistence of coalitions, once formed. Although some research assumes that coalitions are dissolved once their objectives are reached (Drahos, 2003), coalitions in the climate negotiations do not disband easily. This is unsurprising for regional coalitions that are highly institutionalised regional organisations, such as SICA, the Arab Group, or CARICOM, as mentioned earlier. But even climate-specific coalitions tend to persist, and this applies both to formal coalitions such as AILAC and to informal ones such as BASIC or HAC. Few coalitions have become inactive. Members of the Visegrad Group and CG-11 have at least in part joined the EU and hence were dissolved. CACAM was only active over a short period of time, focusing exclusively on the question of the status of its members under the Convention (Yamin & Depledge, 2004). For many of the new coalitions, it remains to be seen to what extent they will continue to be active.

Three factors could explain the persistence of coalitions: first, it takes time for their objectives to be met (if at all) in the climate change negotiations. Even where agreement has been reached, further negotiations refine and interpret this agreement and define rules for its implementation. Frequently, these detailed rules are even more critical than the political agreement per se. The climate change negotiations are an ongoing process, and so the conditions for disbanding have mostly not yet been met. Second, coalition formation requires considerable effort. Once a coalition has successfully been formed and transaction costs have been reduced, members are likely reluctant to discard these efforts (Narlikar, 2003, p. 2). Third, coalition formation often also creates some sort of shared identity and solidarity towards other coalition members. For many observers, this shared identity is the central factor that keeps the extremely diverse G77 together (Ciplet et al., 2015; Vihma et al., 2011).

Of course, coalition persistence does not necessarily mean that the level of activity or coalition membership remain stable over time. A coalition could be very active at one COP, hold daily coordination meetings, and make a large number of joint submissions and statements, but be rather passive at a later COP. Level of activity will depend, among other things, on the level of cohesion and agenda. If coalition members are divided on key agenda items and cannot agree on a common position, the coalition is likely to make fewer joint statements. Similarly, some coalitions coordinate on specific agenda items only; if this item is not on the agenda of a COP, the coalition may not consider coordination necessary (Carter, 2018). Further, coalition membership evolves over time, with new members joining and others leaving a coalition. The EU, for example, grew from 15 members at COP1 to 27 members at present. AOSIS was originally formed by 24 small island states, with other island states joining later (Bouchard, 2004; Ronneberg, 2016).

Implications for the negotiations

The increase in the number of coalitions necessarily leads to multiple and overlapping coalition memberships. Countries do not typically leave one coalition to join another; rather, coalition membership is additive, "creating a multi-layered system of alliances that shapes the political dynamics of the climate change negotiations" (Yamin & Depledge, 2004, p. 33). Particularly non-Annex I countries routinely belong to more than one coalition, combining membership in large global coalitions such as the G77 with membership in smaller, more specific coalitions such as the AGN, CfRN, or AILAC (Bhandary, 2017; Watts & Depledge, 2018). The Democratic Republic of the Congo thus belongs to ten different coalitions, while Sudan is a member of eight different coalitions.[5] Coalitions are of particular relevance to smaller, less powerful parties, and it is therefore not surprising that smaller countries are more likely to join and work through coalitions (Castro & Klöck, 2018; Chasek, 2001; Watts & Depledge, 2018). Many of these have also recently increased their delegation size, potentially to be more active in coalitions (Martinez et al., 2019).

To what extent do these multiple coalition memberships then help or hinder these countries in making their voice heard? How do multiple coalition memberships affect the negotiation process overall? We posit two contrasting effects: mutual support on the one hand, and decreasing influence on the other (see also Klöck, 2020).

On the one hand, smaller and more interest-specific coalitions can complement larger coalitions and lead to a division of labour and mutual support. Chasek and Rajamani (2003, p. 255) conclude that

> multiple memberships in both broad coalitions and small ones appear to confer greater leverage: while the small, issue-focused groups help define, voice, and protect the shared interests of its members, the broad coalitions may offer more general support.

Through members being part of other coalitions, a coalition can forge cross-coalition alliances, ensure the coalition's position is represented in other coalitions, too, and thus garner wider support for its objectives. Loss and damage at the Paris COP illustrate these partnerships and cooperation. AOSIS and LDCs share concerns about loss and damage, and managed to win G77 support: "SIDS in collaboration with LDCs worked tirelessly to ensure that the loss-and-damage concept was incorporated within the Agreement. The first step in this strategy was to get the whole of the G77 and China on board" (Fry, 2016, p. 107). In a second step, the group "agreed on language to attain the consensus from all the smaller coalitions within the global south alliance" (Carter, 2018, p. 235). From this perspective, multiple coalition memberships increase the influence of smaller and poorer countries.

On the other hand, multiple coalition memberships also pose a challenge, particularly for these smaller and poorer countries who only have a few delegates at any one meeting (Benjamin, 2011; Carter, 2018). Negotiations occur in multiple bodies and subgroups at the same time. Carter (2018, p. 84) counts 17 different meetings at one point at the Lima COP – excluding spin-off groups, informal consultations, or side events. Accordingly, it may be "difficult or impossible for smaller delegations to move between coalitions" (Mrema & Ramakrishna, 2010, p. 190). As Chan (2013, p. 277) writes,

> there is a practical and logistical question about the extent to which [a small delegation] can effectively participate in all of these, given delegation constraints, but there is a strategic question too, about how differences in the positions that emerge within different coalitions are reconciled, if at all.

For African countries, Atela et al. (2017, p. 477) therefore conclude that "African delegations also get disfranchised by several coalitions pursuing different interests [...]. The position of the small number of African delegations gets further weakened through the multiple layers of interests and coalitions". Given these logistical and strategic challenges of multiple coalition memberships, it is questionable to what extent the increase in coalitions in the UNFCCC process has helped to level the playing field, notably because smaller countries, for whom the challenges just mentioned are particularly acute, are more likely to belong to many coalitions (Castro & Klöck, 2018).

There is thus anecdotal evidence for both effects – mutual support and decreasing influence. To what extent these effects play out in practice will require additional empirical research.

Conclusion

This chapter seeks to provide an overview of the fragmenting landscape of climate diplomacy. As any multilateral negotiation, the UNFCCC process has always functioned through coalitions. The number of such coalitions active in the climate domain has increased significantly since the conclusion of the 1992 Convention and the 1997 Kyoto Protocol. How can we make sense of this proliferation of climate coalitions, and what are its effects on the negotiations?

We addressed this question by mapping out and systematically describing coalitions that are or were active in the climate change negotiations. We used three dimensions to characterise coalitions: geographic and thematic *scope*; membership *size*; and level of *formality*. While climate coalitions are highly diverse in terms of these dimensions, we identify three distinct clusters of coalitions: (*i*) regional coalitions, which are mostly regional organisations that have started to act as a group in the UNFCCC process; (*ii*) global climate-specific, or instrumental, coalitions that have been created specifically within the climate context; and (*iii*) global generic coalitions that have their origins outside of the climate context. When mapping these coalitions over time, we see that it is particularly global climate-specific coalitions that have been created since the Kyoto Protocol entered into force in 2005 – although some of the regional organisations also became active only in recent years. We also see that coalitions tend to be sticky, or to persist over time. This persistence also implies that coalitions are additive, in the sense that countries join new coalitions, but rarely leave coalitions in which they already are a member. As a result, most countries, in particular poorer and smaller ones, belong to several – we count up to ten – coalitions. We posit two possible – and contrasting – effects of such multiple overlaying coalition memberships: on the one hand, coalitions can mutually support each other, and members that belong to several coalitions can forge new alliances and build bridges across coalitions. On the other hand, there are also logistical challenges of multiple coalition memberships, as well as potential tensions between coalitions that advance divergent positions – even if these coalitions "share" some members.

Our analysis here is but a first step toward a more comprehensive understanding of the changing landscape of climate diplomacy. We focus on the 25 coalitions or groups that have advanced common positions in the negotiations through joint statements, as summarised in the *Earth Negotiations Bulletin* and/or joint submissions. We treat these 25 coalitions in the same way, even though they were created for different purposes and operate in different ways. To what extent these different coalitions can be compared is an open question. In particular, some observers see the Cartagena Dialogue and HAC as distinct from regular coalitions in our sense, because these two groups operated across very many different coalitions, included both Annex I and non-Annex I countries, and coordinated rather loosely, informally, and behind closed doors. Similarly, our approach excludes ad-hoc groups such as the "Green Group" that appeared at COP1 in 1995, but disbanded immediately after their proposal (or "Green Paper") had been adopted as the basis for the Berlin Mandate (Afionis, 2011; Yamin & Depledge, 2004).

Finally, our analysis provides a starting point and toolkit for future research that examines in more depth the effects of coalition characteristics. The dimensions we have used to distinguish between coalitions can be used as explanatory variables, and related to coalition influence, but also their level of activity, longevity, or bargaining strategies, among others. We could, for example, expect the level of formality to increase over time, as coalitions mature and their coordination becomes more routinised and institutionalised. Increasing levels of formality may in turn lead to coalition persistence, and affect the coalition's level of activity and strategies. More formal coalitions could be more – or less – effective. To what extent we find such patterns is an open empirical question. The case studies in part 2 of this volume take on this challenge, and explore in more depth how coalition features can explain coalition formation and effects.

Notes

1 For all coalitions listed on the UNFCCC website, see https://unfccc.int/process-and -meetings/parties-non-party-stakeholders/parties/party-groupings.
2 Some of these groups, particularly the item-specific ones, resemble what Audet (2013) describes as alliance coalitions. However, others (such as AOSIS) have been classified by her as bloc coalitions, given that their interests in the climate negotiations cover more than one topic, and that coalition formation relies not only on strategic interests, but also on a common identity. Again, our classification relies on one single criterion and is therefore more specific than Audet's. In addition, our empirical data fails to show a clear differentiation of coalitions with respect to their duration, which is one aspect used by Audet to differentiate between alliance and bloc coalitions.
3 For the purpose of inferential research, a more granular definition of size (e.g. the actual number of coalition members) might be more appropriate than our simple dichotomous measure. Since our goal is to reach a general classification, this dichotomous measure is more appropriate. In addition, other notions of "size" are conceivable, such as the population covered by a coalition (i.e. the sum of the populations of its members), or its economic size (total GDP covered). Given that in the climate negotiations each country has one voice, we consider the size in terms of number of members to be more pertinent for our purposes. Other measurements may however be more appropriate, for example, when characterizing the power of coalitions.
4 At http://unfccc.int/parties_and_observers/parties/negotiating_groups/items/2714.php or at https://unfccc.int/resource/docs/publications/handbook.pdf.
5 The Democratic Republic of the Congo is a member of AGN, the Cartagena Dialogue, CfRN, COMIFAC, the Congo Basin Group, CVF, G77, HAC, LDCs, and LMDCs; Sudan is a member of AGN, the Arab Group, CfRN, CVF, G77, HAC, LDCs, and LMDCs.

References

Afionis, S. (2011). The European Union as a negotiator in the international climate change regime. *International Environmental Agreements: Politics, Law and Economics*, *11*, 341–360.

Afionis, S. (2017). *The European Union in International Climate Change Negotiations*. London: Routledge.

Atela, J. O., Quinn, C. H., Arhin, A. A., Duguma, L., & Mbeva, K. L. (2017). Exploring the agency of Africa in climate change negotiations: The case of REDD+. *International Environmental Agreements: Politics, Law and Economics*, *17*, 473–482.

Audet, R. (2013). Climate justice and bargaining coalitions: A discourse analysis. *International Environmental Agreements: Politics, Law and Economics*, *13*, 369–386.

Benjamin, L. (2011). The Role of the Alliance of Small Island States (AOSIS) in UNFCCC Negotiations. In E. Couzens & T. Honkonen (Eds.), *International Environmental Lawmaking and Diplomacy Review 2010* (pp. 117–132). Joensuu: University of Eastern Finland.

Betzold, C. (2010). 'Borrowing' power to influence international negotiations: AOSIS in the climate change regime, 1990–1997. *Politics*, *30*(3), 131–148.

Betzold, C., Castro, P., & Weiler, F. (2012). AOSIS in the UNFCCC negotiations: From unity to fragmentation? *Climate Policy*, *12*(5), 591–613.

Bhandary, R. R. (2017). Coalition strategies in the climate negotiations: An analysis of mountain-related coalitions. *International Environmental Agreements: Politics, Law and Economics*, *17*(2), 173–190.

Blaxekjær, L. Ø., & Nielsen, T. D. (2015). Mapping the narrative positions of new political groups under the UNFCCC. *Climate Policy*, *15*(6), 751–766.

Bouchard, C. (2004). Les petits états et territoires insulaires. In *Document n° 51*. Saint-Denis: Observatoire du développement de La Réunion.

Brütsch, C., & Papa, M. (2013). Deconstructing the BRICS: Bargaining coalition, imagined community, or geopolitical fad? *The Chinese Journal of International Politics*, *6*, 299–327.

Carter, G. (2015). Establishing a Pacific Voice in the Climate Change Negotiations. In G. Fry & S. Tarte (Eds.), *The New Pacific Diplomacy* (pp. 205–220). Canberra: ANU Press.

Carter, G. (2018). *Multilateral consensus decision making: How Pacific island states build and reach consensus in climate change negotiations* (Doctor of Philosophy). Australian National University, Canberra.

Castro, P., & Klöck, C. (2018). Coalitions in global climate change negotiations. In *INOGOV Policy Brief N°5*. INOGOV: Innovations in Climate Governance.

Cepaluni, G., Galdino, M., & de Oliveira, A. J. (2012). The bigger, the better: Coalitions in the GATT/WTO. *Brazilian Political Science Review*, *6*(2), 28–55.

Chan, N. (2013). *The construction of the South: Developing countries, coalition formation and the UN climate change negotiations, 1988–2012* (DPhil). University of Oxford, Oxford.

Chan, N. (2020). The Temporal Emergence of Developing Country Coalitions. In C. Klöck, P. Castro, F. Weiler, & L. Ø. Blaxekjær (Eds.), *Coalitions in the Climate Change Negotiations*. Abingdon: Routledge.

Chasek, P. (2001). *Earth Negotiations: Analyzing Thirty Years of Environmental Diplomacy*. Tokyo: United Nations University Press.

Chasek, P. (2015). Negotiating Coalitions. In J.-F. Morin & A. Orsini (Eds.), *Essential Concepts of Global Environmental Governance* (pp. 130–134). New York, NY: Routledge.

Chasek, P., & Rajamani, L. (2003). Steps Toward Enhanced Parity: Negotiating Capacities and Strategies of Developing Countries. In I. Kaul, P. Conceiçao, K. Le Goulven, & R. U. Mendoza (Eds.), *Providing Global Public Goods: Managing Globalization* (pp. 245–262). Oxford: Oxford University Press.

Ciplet, D., Khan, M., & Roberts, J. T. (2015). *Power in a Warming World: The New Global Politics of Climate Change and the Remaking of Environmental Inequality*. Boston: MIT Press.

Costantini, V., Crescenzi, R., Filippis, F. D., & Salvatici, L. (2007). Bargaining coalitions in the WTO agricultural negotiations. *The World Economy*, *30*(5), 863–891.

Costantini, V., Sforna, G., & Zoli, M. (2016). Interpreting bargaining strategies of developing countries in climate negotiations: A quantitative approach. *Ecological Economics*, *121*, 128–139.

de Águeda Corneloup, I., & Mol, A. P. J. (2014). Small island developing states and international climate change negotiations: The power of moral "leadership". *International Environmental Agreements: Politics, Law and Economics*, *14*(3), 281–297.

Downie, C., & Williams, M. (2018). After the Paris agreement: What role for the BRICS in global climate governance? *Global Policy*, *9*(3), 398–407.

Drahos, P. (2003). When the weak bargain with the strong: Negotiations in the World Trade Organization. *International Negotiation*, *8*, 79–109.

Dupont, C. (1996). Negotiation as coalition building. *International Negotiation*, *1*(1), 47–64.

Edwards, G., & Roberts, J. T. (2015). *Fragmented Continent: Latin America and the Global Politics of Climate Change*. Cambridge, MA: MIT Press.

Fry, I. (2016). The Paris agreement: An insider's perspective—The role of small island developing states. *Environmental Policy and Law*, *46*(2), 105–108.

Groen, L. (2020). Group Interaction in the UN Framework Convention on Climate Change. In K. E. Smith & K. V. Laatikainen (Eds.), *Group Politics in UN Multilateralism* (pp. 267–284). Leiden and Boston: Brill Nijhoff.

Groen, L., & Niemann, A. (2013). The European Union at the Copenhagen climate negotiations: A case of contested EU actorness and effectiveness. *International Relations*, *27*(3), 308–324.

Hallding, K., Jürisoo, M., Carson, M., & Atteridge, A. (2013). Rising powers: The evolving role of BASIC countries. *Climate Policy*, *13*(5), 608–631.

Hallding, K., Olsson, M., Atteridge, A., Vihma, A., Carson, M., & Román, M. (2011). Together Alone: BASIC Countries and the Climate Change Conundrum. In *TemaNord 2011:530*. Copenhagen: Nordic Council of Ministers.

Hampson, F. O., & Reid, H. (2003). Coalition diversity and normative legitimacy in human security negotiations. *International Negotiation*, *8*, 7–42.

Hirsch, T. (2016). *The Role of Alliances in International Climate Policy After Paris*. Bonn: Friedrich Ebert Stiftung.

International Institute for Sustainable Development. (1997–2018). Earth negotiations bulletin. Retrieved from http://enb.iisd.org/enb

Jernnäs, M., & Linnér, B.-O. (2019). A discursive cartography of nationally determined contributions to the Paris climate agreement. *Global Environmental Change*, *55*, 73–83.

Kameri-Mbote, P. (2016). The Least Developed Countries and Climate Change Law. In K. R. Gray, R. Tarasofsky, & C. Carlarne (Eds.), *The Oxford Handbook of International Climate Change Law* (pp. 740–760). Oxford: Oxford University Press.

Kasa, S., Gullberg, A. T., & Heggelund, G. (2008). The group of 77 in the international climate negotiations: Recent developments and future directions. *International Environmental Agreements: Politics, Law and Economics*, *8*(2), 113–127.

Klöck, C. (2020). Multiple coalition memberships: Helping or hindering small states in multilateral (climate) negotiations? *International Negotiation*, 25(2), 279–297.

Laatikainen, K. V. (2017). Conceptualizing groups in UN multilateralism: The diplomatic practice of group politics. *The Hague Journal of Diplomacy*, 12, 113–137.

Laatikainen, K. V., & Smith, K. E. (2020a). Conclusion: "The Only Sin at the UN Is Being Isolated". In K. E. Smith & K. V. Laatikainen (Eds.), *Group Politics in UN Multilateralism* (pp. 305–322). Leiden and Boston: Brill Nijhoff.

Laatikainen, K. V., & Smith, K. E. (2020b). Introduction: Group Politics in UN Multilateralism. In K. E. Smith & K. V. Laatikainen (Eds.), *Group Politics in UN Multilateralism* (pp. 3–19). Leiden and Boston: Brill Nijhoff.

Martinez, G. S., Hansen, J. I., Holm Olsen, K., Ackom, E. K., Haselip, J. A., Bois von Kursk, O., & Bekker-Nielsen Dunbar, M. (2019). Delegation size and equity in climate negotiations: An exploration of key issues. *Carbon Management*, 10(4), 431–435.

Mrema, E., & Ramakrishna, K. (2010). The Importance of Alliances, Groups and Partnerships in International Environmental Negotiations. In T. Honkonen & E. Couzens (Eds.), *International Environmental Law-Making and Diplomacy Review 2009* (pp. 183–192). Joensuu: University of Eastern Finland.

Muldoon, J. (2005). *Multilateral Diplomacy and the United Nations Today*. Boulder, CO: Westview.

Narlikar, A. (2003). *International Trade and Developing Countries: Bargaining Coalitions in the GATT & WTO*. London and New York: Routledge.

Odell, J. S. (2006). *Negotiating Trade: Developing Countries in the WTO and NAFTA*. Cambridge: Cambridge University Press.

Ott, H. E., Bauer, S., Brandi, C., Mersmann, F., & Weischer, L. (2016). Climate Alliances après Paris: The Potential of Pioneer Climate Alliances to Contribute to Stronger Mitigation and Transformation.

Panke, D. (2013). *Unequal Actors in Equalising Institutions: Negotiations in the United Nations General Assembly*. Basingstoke: Palgrave Macmillan.

Richards, M. (2001). *A Review of the Effectiveness of Developing Country Participation in the Climate Change Convention Negotiations*. London: Overseas Development Institute.

Roger, C. (2013). African Enfranchisement in Global Climate Change Negotiations. In *Africa Portal Backgrounder* (Vol. 57, pp. 1–10). Waterloo, Canada: Centre for International Governance Innovation.

Ronneberg, E. (2016). Small Islands and the Big Issue: Climate Change and the Role of the Alliance of Small Island States. In K. R. Gray, R. Tarasofsky, & C. Carlarne (Eds.), *The Oxford Handbook of International Climate Change Law* (pp. 761–778). Oxford: Oxford University Press.

Starkey, B., Boyer, M. A., & Wilkenfeld, J. (2005). *Negotiating a Complex World: An Introduction to International Negotiation* (2nd ed.). Lanham, MD: Rowman & Littlefield.

Stephenson, S. R., Oculi, N., Bauer, A., & Carhuayano, S. (2019). Convergence and divergence of UNFCCC nationally determined contributions. *Annals of the American Association of Geographers*, 109(4), 1240–1261.

Tobin, P., Schmidt, N. M., Tosun, J., & Burns, C. (2018). Mapping states' Paris climate pledges: Analysing targets and groups at COP21. *Global Environmental Change*, 48, 11–21.

Vihma, A., Mulugetta, Y., & Karlsson-Vinkhuyzen, S. (2011). Negotiating solidarity? The G77 through the prism of climate change negotiations. *Global Change, Peace & Security*, 23(3), 315–334.

Wagner, L. M. (1999). Negotiations in the UN commission on sustainable development: Coalitions, processes, and outcomes. *International Negotiation*, *4*(1), 107–131.

Watts, J., & Depledge, J. (2018). Latin America in the climate change negotiations: Exploring the AILAC and ALBA coalitions. *Wiley Interdisciplinary Reviews: Climate Change*, *9*(6), e533. doi:10.1002/wcc.533

Weiler, F., & Castro, P. (2020). "Necessity Has Made Us allies": The Role of Coalitions in the Climate Change Negotiations. In C. Klöck, P. Castro, F. Weiler, & L. Ø. Blaxekjær (Eds.), *Coalitions in the Climate Change Negotiations*. Abingdon: Routledge.

Williams, M. (2005). The third world and global environmental negotiations: Interests, institutions and ideas. *Global Environmental Politics*, *5*(3), 48–69.

Woods, B. A., & Kristófersson, D. M. (2016). The state of coalitions in international climate change negotiations and implications for global climate policy. *International Journal of Environmental Policy and Decision Making*, *2*(1), 41–68.

Yamin, F., & Depledge, J. (2004). *The International Climate Change Regime: A Guide to Rules, Institutions and Procedures*. Cambridge: Cambridge University Press.

3 "Necessity has made us allies"

The role of coalitions in the climate change negotiations

Florian Weiler and Paula Castro

Introduction

As described in the previous chapters, the multilateral climate change negotiations under the United Nations Framework Convention on Climate Change (UNFCCC) are structured around coalitions – groups of countries that voluntarily get together in order to put forward common positions in the negotiations, thereby gaining visibility and strength in numbers (Klöck, Weiler, & Castro, 2020). Given the increasing number of frequently overlapping coalitions over time and their wide diversity in terms of size, geographic scope, thematic focus, and other characteristics, a key question is which role coalitions actually play in the negotiations. This question has, of course, been addressed before. Generally, such studies conclude that coalitions help to narrow the bargaining space (e.g. Zartman, 1994), particularly in complex multi-party and multi-issue negotiations such as the climate negotiations (Muldoon, 2018). This facilitates finding a negotiated agreement as the number of actors is reduced (Dupont, 1994, 1996).

In this study we further this literature by focusing on whether *differences between coalitions lead to differences in how closely they cooperate, and therefore to differences in how they participate in the negotiations.* We base our analysis on the coalition types defined by Castro and Klöck (2020), and on the assumption that coalitions of (theoretically) more aligned parties play a more central role in the negotiations. This is based on the idea that countries can pool power with other negotiation partners (Henry, 2011; Weible, 2005). This pooling of power is easier when the preferences of countries are linked and closely reflect each other. If they can find common positions, it is therefore advantageous for coalitions to work as closely together as possible, and thereby increase each member's power. The more common positions they can find, the more the coalitions will come to the fore and play a more central role, while their members individually should play lesser roles.

While this pooling of interests explains – at least in part – why coalitions form in the first place, it tells us relatively little about how coalitions may differ, for example, in terms of their levels of participation in the negotiations, and why. We address this research gap by drawing on the Advocacy Coalition Framework and on Resource Dependence Theory to derive hypotheses regarding which types of coalitions play a more central role at the negotiations. Our empirical test relies on an understanding of the climate negotiations as networks of negotiation interactions between countries and coalitions. In these networks, countries and coalitions are the actors, and the ties between them are their cooperative or conflictive negotiation exchanges over time. We therefore construct annual networks of cooperative negotiation behaviour between countries and coalitions, and test our hypotheses by analysing the structural roles adopted by coalitions in these networks. Specifically, we study different types of network centrality (see e.g. Freeman, 1979). Which coalitions tend to be more active in oral exchanges during the negotiations (out-degree)? Which coalitions are more popular, and therefore more frequently supported in their position statements by other groups and countries (in-degree)? Which ones work as brokers, thereby connecting actors with different views and positions (betweenness and closeness)?

To obtain these centrality measures, we rely on an existing dataset on countries' and coalitions' behaviour at the oral UNFCCC negotiations between 1995 and 2013 (Castro, 2017). The data codes the summaries of the negotiations published in the *Earth Negotiations Bulletin* (IISD, 1995–2013) to record all reported instances of cooperative or conflictive negotiation behaviour between pairs of countries or coalitions.[1] We apply regression analysis to test whether different coalition types – based on the categorisation proposed by Castro and Klöck (2020) – play a more or less central role in the negotiations. We find that more regionally focused coalitions (as opposed to coalitions with global reach), climate-specific coalitions (as opposed to generic ones), and larger coalitions are more central on all four centrality measures, and thus play a more important role in the climate negotiations.

In the next section we introduce our theoretical expectations for the various types of coalitions. We then explain how our cooperation networks were coded and describe the methodology used in this study. Then we interpret the results of our descriptive and inferential analyses, and the last section concludes.

Theoretical background

Coalitions are central to the study of complex multi-party, multi-issue negotiations such as climate change, since they reduce complexity and facilitate finding an agreement (Dupont, 1994, 1996; Hopmann, 1996; Zartman, 1994). According to Olson (1965), some form of coordination among the various participants is necessary to achieve cooperative outcomes in such complex settings. Dillenbourg et al. (1995) define coordination as an effort by all or a subset of countries involved in negotiations to act in unity in the pursuit of a common goal. This common goal is to achieve an agreement as close to their initial position, and thus as favourable to them, as possible. Coordination with others is thus a crucial part of a country's bargaining strategy. Such coordination may be formalised in the form of established coalitions, or it may be more short-term and focused on specific negotiation issues (Klöck et al., 2020; Ott, Bauer, Brandi, Mersmann, & Weischer, 2016).

We apply a theoretical framework frequently used in studies of policy change and policy networks at the national or local level to conceptualise the reasons why particular sets of countries decide to cooperate and coordinate their efforts by building bargaining coalitions, and the role that those coalitions play in the negotiations. The Advocacy Coalition Framework (AFC, see e.g. Sabatier & Jenkins-Smith, 1993) asserts that cooperation mostly emerges among actors that share similar policy beliefs through a mechanism called "biased assimilation". This assumes that actors "tend to interpret evidence in a way that supports their prior beliefs and values" (Henry, 2011, p. 365). Therefore, actors with similar prior beliefs will interpret the same policy information in a similar way, which leads to trust and eventually to coordination and coalition building. Coalition formation thus builds on the "tendency for [countries] to form ties based on common attributes ... to share strength and minimize weaknesses" (Hafner-Burton, Kahler, & Montgomery, 2009, pp. 567–568). In network terminology this is also called homophily, i.e. the tendency of actors with similar characteristics to "flock together" (Goodreau, Kitts, & Morris, 2009).

The Resource Dependence Theory (RDT) posits that actors in negotiations can bundle their resources through cooperation, and thus increase their (perceived) power and influence over the negotiation process (Henry, 2011; Weible, 2005). Most individual countries involved in the climate negotiations lack the power to influence the outcome of the talks alone.[2] However, when aligning and coordinating with other negotiation parties, for instance by acting as coalitions, individual actors can increase their power by pooling their resources with (stronger) negotiating partners, and thus increase their chances of success (Bailer, 2004; Rubin & Zartman, 2000; Zartman, 1997). Ashe, van Lierop, & Cherian (1999) and Betzold (2010) have shown that the Alliance of Small Island States (AOSIS) was able to use such tactics in the climate negotiations between 1990 and 1997 with remarkable success.

We thus set out from the assumption that due to biased assimilation and the possibility of pooling power with each other, countries start to build coalitions and coordinate negotiation positions – at least on some topics discussed in the negotiations. Building upon this assumption, we propose two pathways for how coalitions can become influential in the negotiations. First, based on the ACF, we propose that countries work more closely together the more like-minded (i.e. similar) they are, particularly with respect to the topics that are being negotiated. While coalitions are built on the basis of similarities of positions, coalition members do not necessarily hold similar views on all climate-related issues being discussed (see e.g. Tobin, Schmidt, Tosun, & Burns, 2018). For example, coalitions that are built on the basis of pre-existing political alliances (such as the Group of 77 and China, G77) or of other non-climate-related common characteristics (such as development status for the Least Developed Countries, LDCs) might have substantial differences regarding specific climate-related concerns (see e.g. DeSombre, 2000). Hence, the more similar members of a coalition are with respect to their climate-related interests, the better – we argue – they will be able to reach common positions. Coalitions that "work better" in this sense will, in turn, play a larger role in the negotiations, because countries will rely more on the group to make their voice heard instead of trying to influence the negotiations on their own. In sum, more cohesive coalitions are expected to play a more central role in the climate negotiations (see also Ciplet, Roberts, & Khan, 2015; Panke, 2013).

The second pathway for coalitions to become influential is through the RDT, i.e. the amount of resources a given coalition is able to bundle. More resources, in this framework, translate into more power. One crucial resource, in this context, is voice. A coalition with a larger membership represents the voices of many parties, and can thus draw strength from sheer numbers. Another resource is financial. Richer and larger countries are better able to bear the costs of dedicated climate specialists and of larger negotiation delegations (Drahos, 2003; Weiler, 2012). Coalitions that bring together such countries will thus on average be better able to keep up with the pace of the negotiations, prepare statements, and speak up more frequently. We thus expect more powerful coalitions to play a larger role at the negotiations.

It should be noted that there might be a trade-off between both causal mechanisms. The more members a coalition has, the more bargaining power it bundles. But at the same time, the larger a coalition, the more likely it is that its members are more heterogeneous with respect to their climate-related preferences (Hamilton & Whalley, 1989; Narlikar, 2004).

How to make sense of these expectations for the increasing number of coalitions active in the climate negotiations? In order to be able to test our expectations for various types of coalitions, we build on Castro and Klöck (2020), who have classified the existing coalitions in the climate negotiations along four dimensions – geographic and thematic scope, size of the coalition, and level of formality. We derive specific expectations for each of these coalition types through the prism of the ACF and RDT frameworks.

Geographic scope: this dimension assesses whether the members of a coalition are located within the same geographic region (regional coalition) or not (global coalition). Countries belonging to the same region are likely to face similar problems caused by climate change (Rosenzweig et al., 2008)[3], and are more likely to share a similar political culture and thus a similar approach to policymaking. We would thus expect countries within the same region to share more similar climate-related policy beliefs than countries on a global scale. The ACF would thus suggest that regional coalitions should be more cohesive than global ones. All else being equal (including coalition size and resources), we therefore hypothesise that regional coalitions play a more central role in the climate negotiations than global ones.

Thematic scope: this dimension refers to the fact that some coalitions were created purposefully with a focus on climate change (climate-specific coalitions), while others originated in other international forums and thus work also on other issues beyond climate change (generic coalitions). Since climate-specific coalitions were created specifically in the context of the UNFCCC negotiations, this is the topic in which the most coordination among their members has taken place. In contrast, generic coalitions may share commonalities in other areas such as trade, while their climate-related interests might differ more. The ACF therefore suggests closer cooperation and stronger cohesion within climate-specific coalitions. Thus, we expect climate-specific coalitions to play a more central role in the climate negotiations than coalitions with a generic scope. A well-known example for a thematic coalition able to play a central role in the climate negotiations is AOSIS (Betzold, Castro, & Weiler, 2012).

Coalition size: the larger coalitions are, according to the RDT, the more power should they be able to accumulate (see Panke, 2013; Weiler, 2012), holding other things like geographic and thematic scope constant. The reasons for pooling power are twofold. First, countries can benefit from each other's external power sources e.g. total GDP or, more specific to the climate negotiation, greenhouse gas emissions, and thus increase their combined weight in the negotiations (Krasner, 1991; Milner, 1992). Second, they can also share their internal power sources by linking up their delegations and respective skill sets (Antonides, 1991; Snyder & Diesing, 2015). Thus, we expect larger coalitions to play a more central role in the climate negotiations than smaller ones.

Level of formality: the level of formality is about how much internal structure coalitions have. More formal coalitions have official founding documents and develop procedures on how to cooperate and how to find internal agreement; some even have official structures, such as secretariats, with related staff and resources, while for more informal coalitions cooperation is much looser and often only temporary (Drahos, 2003). Based on the RDT, we posit that formal coalitions have a resource-related advantage over more ad-hoc, informal negotiating groups, and are therefore more central in the climate negotiations. The reason behind this is that formalised and more permanent structures enable and facilitate coordination among coalition members. In other words, these structures enable group members to more easily share information, but also to come together and to formulate joint positions, while more informal coalitions rely much more on personal ties among negotiators, and thus on ad-hoc coordination during negotiation rounds.

Research design

In order to test our expectations, we start by presenting some descriptive statistics and graphs of the role that coalitions have historically played in the UNFCCC negotiations, before running panel data regressions.

Data

During the annual UNFCCC negotiation meetings, countries and coalitions have the opportunity to voice their views and positions in various open and closed sessions. Even though there are no official transcripts of these meetings, they have regularly been registered, summarised, and published by the International Institute for Sustainable Development (IISD) in its *Earth Negotiations Bulletin* (ENB) since 1995 (see IISD, 1995–2013). An issue of the ENBs is usually published for every day of negotiations, and summarises the most important statements made by the negotiation parties in the publicly accessible meetings. Even though they consist only of summaries, and they only cover the open meetings, the ENBs are the most authoritative, consistent, and objective accounts of the climate change negotiations, and are used by negotiators, experts, and researchers to inform themselves about the process. Venturini et al. (2014, p. 2) have called the ENBs "a workable proxy with which to analyse all the climate negotiations".

The analysis is based on a dataset of countries' (and coalitions') negotiation exchanges during the UNFCCC meetings, hand-coded from the ENBs (Castro, 2017). All the relevant ENBs during the 19-year period from the first Conference of the Parties (COP1) in Berlin (March to April 1995) to COP19 in Warsaw (November 2013) were coded. The dataset covers 62,097 negotiation interactions among 222 participants (including both individual countries and coalitions) to the UNFCCC negotiations over a total of 85 negotiation rounds (or about 461 negotiation days).[4]

The dataset contains dyadic data on how participants to the UNFCCC negotiations react to other participants' oral interventions during the discussions (Castro, 2017). Each observation represents a negotiation interaction between a pair of countries or coalitions, and records: (i) the actors involved; (ii) the topic of the discussion (e.g. mitigation, adaptation, finance, carbon markets, etc.); (iii) the date in which it took place; and (iv) the type of negotiation interaction. Negotiation interactions can be *cooperative* (country/coalition 1 speaks on behalf of, supports or agrees with country/coalition 2), or *conflictive* (country/coalition 1 criticises, opposes, or attempts to delay a statement or proposal by country/coalition 2).

For example, ENB No. 324 includes following sentence: "On the issue of bunker fuels, SAUDI ARABIA, opposed by NORWAY, the EU and AOSIS, proposed deleting this agenda item". From this sentence, the dataset records three conflictive and three cooperative negotiation interactions: opposition by Norway against Saudi Arabia, opposition by the EU against Saudi Arabia, opposition by AOSIS against Saudi Arabia, agreement between Norway and the EU, agreement between Norway and AOSIS, and agreement between the EU and AOSIS. The topic of the negotiation interactions is coded as "international transport", and the interactions took place on 8 May 2007.[5]

After all such interactions for each negotiation day were coded, they were aggregated for every year in order to obtain a variable counting the number of cooperative negotiation interactions between each pair of actors during that year, and another one counting the number of conflictive interactions. We use the annual count of cooperative interactions to build 19 annual negotiation networks, from which the dependent variables used in this paper are derived.[6]

These dependent variables are four measures of network centrality, and we consider higher centrality values to indicate importance in the network structure.

Specifically, we use the measures in-degree, out-degree, betweenness, and closeness (see Freeman, 1979; Wasserman & Faust, 1994 for how these measures are constructed). The first two centrality measures provide an idea of how well connected a member of the network is. In-degree centrality is based on the number of incoming ties an actor (or node in the network) receives (i.e. how often country A has been supported by its peers). In our case, it is an indicator of the popularity of a country as a target of cooperative negotiation behaviour by its peers. Out-degree centrality is estimated on the basis of the outgoing ties of an actor. It indicates how active a country is in supporting or agreeing with its peers during the negotiations. Betweenness centrality assesses the extent to which an actor lies on the network paths between other actors. Nodes with higher betweenness centrality may have larger influence within the network as they tend to connect many other nodes; betweenness is therefore used to assess structural power within the network. Finally, closeness centrality indicates how close an actor is, on average, to all other actors in the network, and also provides an indication of that actor's level of connectedness.

We use the *igraph* package in R (Csardi & Nepusz, 2006) to calculate these four quantities for all coalitions and countries present in the networks in a given year. Since we are only interested in how central the coalitions are, we only keep the values for those coalitions that officially existed in a given year. For instance, in the first two years for which we have data, only five coalitions actively participated in the negotiations (G77, EU, AOSIS, African Group of Negotiators (AGN), OPEC), while in the last two years 18 groups are included in the dataset. Overall, the dataset includes 200 observations for 21 coalitions. Five of these coalitions are present in all 19 years covered by the dataset (AOSIS, LDCs, AGN, G77, EU), while many groups only emerge later.

We complement this data with information on the coalitions, including their thematic and geographic scope, membership size, and formality obtained from Castro and Klöck (2020). This information constitutes our independent variables as described below.

Geographic scope: classifies coalitions into regional and global groups, with 11
    coalitions falling into the former category, and ten into the latter.
Thematic scope: nine of the coalitions in the dataset are climate-specific, while 12
    of them are generic.
Coalition size: this is coded differently from what Castro and Klöck (2020) propose. While they simply classify coalitions as being large or small, we capture
    the size more accurately by including the number of active coalition members. The minimum value is three (Mountainous Landlocked Developing
    Countries), and the maximum is 129 (G77).[7]
Level of formality: this is about whether coalitions have formal structures in
    place, or whether they entail more ad-hoc cooperation. 14 of our coalitions
    have been classified as formal, and seven as informal.

In addition to these four variables that reflect our hypotheses, we also include a variable capturing for how many years over the time-horizon of the study a group has been active.[8]

Descriptive analysis

To present an overview of the networks of cooperative interactions, we generate network graphs for four negotiation periods: 1995 to 2001 (negotiations of the Kyoto Protocol ending with the Marrakech Accords), 2002 to 2006 (Kyoto implementation phase), 2007 to 2009 (Bali Action Plan and attempt to reach a post-2012 deal), 2010 to 2013 (recovery after COP 15 and start of negotiations on post-2020 deal). To create these aggregate graphs, we add the respective annual networks, meaning that if two actors during the negotiation period cooperated once in two of the years, their final value in the combined network will be two. However, as the aggregate networks would be too dense for graphical representation, the four graphs in Figure 3.1 only show dyads with an average of at least two ties per year. For instance, in the first negotiation period, which covers seven years, only those dyads that had at least 14 cooperative negotiation interactions over this period of time are shown to have a tie. In addition, we implement a hierarchical agglomeration algorithm able to detect community structures within networks (Clauset, Newman, & Moore, 2004); in other words, an algorithm that automatically detects clusters within the four negotiation periods. Note that this clustering is calculated separately for the four graphs, meaning that a) clusters in different periods do not influence each other, and b) different symbols may be used for similar clusters across time periods. To highlight the coalitions, we have plotted their nodes slightly larger than those of individual countries.

Regression analysis

Based on the annual dataset, we apply conventional panel regression models to check whether differently categorised coalitions exhibit, as hypothesised, higher centrality values (and are thus more important) in the networks. In order to control for the (unbalanced) panel structure – particularly in the first years, we have only few observations, but this value increases strongly in later years – we also include yearly fixed effects in the models.

Results

Before we discuss the regression results testing our hypotheses, we take a descriptive look at the networks on which the analysis is based, and the various centrality measures we obtained for the negotiation coalitions.

Descriptive analysis

Figure 3.1 shows the network graphs for the four negotiation periods described above. Those negotiation coalitions which appear in the networks are frequently central players, and tend to be surrounded (unsurprisingly) by their members, but also by other coalitions. In the first two periods only three and four coalitions are represented, respectively (other coalitions either did not exist or were not active enough), i.e. the EU, AOSIS, G77, and, in 2002–2006, the AGN. In the

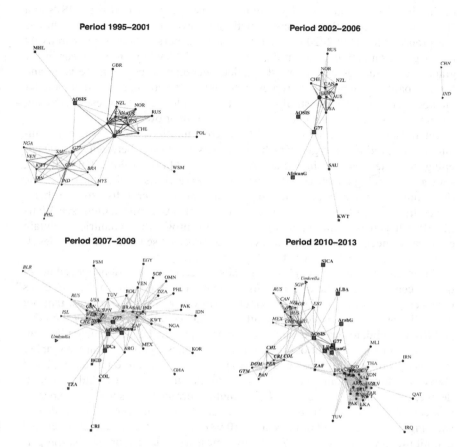

Figure 3.1 Networks of the four negotiation periods. A tie means there were on average at least two cooperative negotiation interactions per year between the pair of actors. Different shapes and font styles indicate the communities identified by the agglomeration algorithm. Darker lines indicate stronger cooperation ties.

period 2007–2009 also, the LDCs and the Umbrella Group become more active, but of these, only the LDCs seem to play a relatively central role. From 2010 onwards, several other coalitions become active in the negotiations. Of them, only the Environmental Integrity Group (EIG) adopts a central position in the network. In addition, several of the new coalitions identified by Castro and Klöck (2020) fail to appear in any of the graphs. They do not seem to participate very often in the official UNFCCC negotiations, and may thus rather act informally on the side of the negotiation process (see for example, Blaxekjær (2020) on the Cartagena Dialogue).

In all periods, the EU is surrounded by and forms a community with other developed countries and developed country coalitions (EIG and Umbrella Group).

In the first two periods, the developing country coalitions (led by AOSIS and G77) also tend to form communities with their own members. This tendency is less pronounced in 2002–2006, because during this period there is comparatively less negotiation activity than in the others, so only few countries appear in the graph. This indicates that during this quieter negotiation period these large and formal coalitions stepped in to represent the interests particularly of their smaller members vis-à-vis the industrialised countries (while the two largest developing countries, India and China, even built a cluster of their own).

Interestingly, from 2007 onwards, these communities start to change: a community of coalitions emerges, which clearly acts as a bridge between most developing countries and the industrialised ones. In addition, from 2010 onwards, several developing countries with more progressive views on the climate regime (including Chile, Colombia, and Costa Rica, among others) disassociate themselves from the bulk of developing countries (for more information, see Watts (2020), Chapter 9 in this volume). While these progressive countries maintain cooperative negotiation interactions with the industrialised ones, all other developing countries now have very few direct ties with them.

All in all, and unsurprisingly, the graphs suggest that coalitions indeed adopt quite a central role in the negotiations, building a bridge between the two blocs of developing and industrialised countries. However, they also suggest that not all coalitions are equally central. While the "old" coalitions, including the G77, AOSIS, EU, and, to certain extent, also the AGN and the LDCs, seem to play central roles, most other coalitions appear rather peripherally, if at all, in the graphs.

Exactly how central these coalitions are, and why, is the question proposed above. The four centrality measures we use to test our hypotheses (in-degree, out-degree, betweenness, and closeness) are highly correlated, as should be expected (Valente, Coronges, Lakon, & Costenbader, 2008), with in-degree and out-degree showing the highest level of correlation (0.98), while closeness and betweenness exhibit the lowest level (0.37). As the four measures thus capture similar properties of the networks, we use out-degree to exemplify how the centrality of the coalitions varies over time. Figure 3.2 represents this graphically. As there are many coalitions, we split them in order to avoid over-crowding in the graph. In the top panel a), out-degree is depicted for "old" coalitions, i.e. coalitions that were established before 2007 and are still active. Panel b) shows the same measure for newer coalitions (established 2007 or later), and groups that are no longer active. Finally, in panel c), we show overall activity levels of coalitions in the negotiations in the different years, first by looking at total number of statements made by all coalitions (black line), and second by the number of coalitions active in a given year (grey line).

As can be seen, in the first two years the (few) active groups are relatively central, and their overall activity in terms of statements made in the negotiations is relatively high (769 and 698 statements for 1995 and 1996). In the following years, both their activity level and their centrality tend to drop, with the exception of the EU, which remains the most active group throughout. After lingering at lower activity levels for quite some time, around 2007, coalitions tend to become

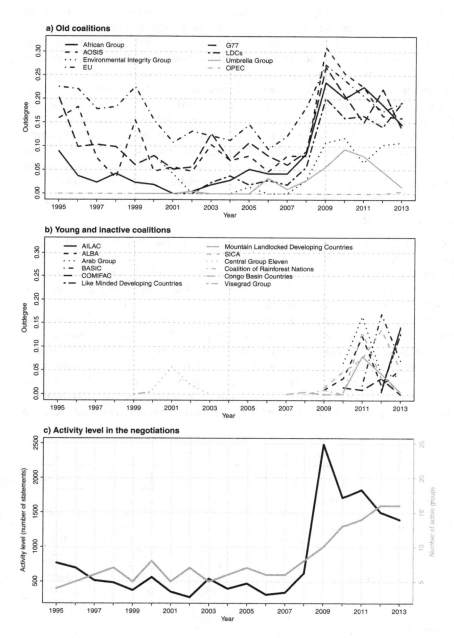

Figure 3.2 Centrality (out-degree) of old (panel a) and young (panel b) coalitions, and overall activity levels of coalitions (panel c) between 1995 and 2013.

more active again, reaching almost 2500 statements overall in 2009, a year in which many of the active groups also register very high network centrality values. Thereafter, overall activity remains high, yet drops slightly from the peak in 2009, as do the centrality scores of many of the coalitions. This is the period in which many new coalitions come into the picture. Interestingly, many of these new groups reach a peak of centrality in the second or third year of their existence, but this value often drops again in the year thereafter. This might indicate that the expectations of their members are initially high, but seem to decrease after a while, so that less time and fewer resources are spent on coordinating and putting forward common positions. While interesting, this inspection of the data allows little inference regarding the proposed hypotheses, which is why we now turn to the regression results on how coalitions characteristics influence centrality.

Regression results

Table 3.1 and Figure 3.3 show the regression results for all four centrality measures derived from the networks. Due to the high correlation between the four dependent variables, the results in the four models are highly similar. In both the table and figure, for the five dichotomous variables we indicate the category to which the coefficient applies in brackets.

The *geographic scope* variable is positive and statistically significant in all four models, meaning that coalitions with a regional focus tend to be more central in the negotiation networks than coalitions with members from all over the world. This is in line with our expectation, as we argued that regional groups share more similar climate change concerns than more geographically dispersed groups, and should thus be more cohesive, leading to more active participation in the negotiations. The magnitude of the effect of being a regional coalition represents between

Table 3.1 Results of the four centrality models (standard errors in parentheses)

	In-Degree	Out-Degree	Betweenness	Closeness
Geographic scope (regional)	0.053***	0.054***	0.366***	0.110***
	(0.012)	(0.012)	(0.101)	(0.026)
Thematic scope (generic)	−0.061***	−0.061***	−0.311**	−0.165***
	(0.014)	(0.014)	(0.116)	(0.030)
Coalition size	0.001***	0.001***	0.006***	0.002***
	(0.0002)	(0.0001)	(0.001)	(0.0003)
Level of formality (informal)	−0.014	−0.018	−0.124	−0.025
	(0.014)	(0.014)	(0.121)	(0.031)
Years active	0.040***	0.041***	0.246***	−0.006
	(0.008)	(0.008)	(0.070)	(0.018)
(Intercept)	−0.004	−0.007	−0.510	0.777***
	(0.038)	(0.037)	(0.318)	(0.082)
N	200	200	200	200
R²	0.49	0.45	0.31	0.57

*** $p < 0.001$; ** $p < 0.01$; * $p < 0.05$, year fixed effects included.

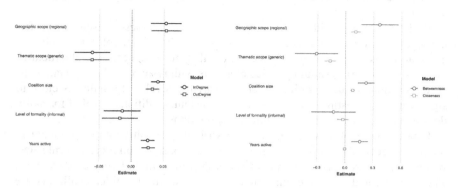

Figure 3.3 Graphical representation of the results of the four centrality models (including 95% confidence intervals).

14% and 18% of the total range of the respective dependent variable in all four models, and is thus not only statistically significant, but also substantive. We thus conclude that our geographic scope hypothesis is substantiated.

Turning to the *thematic scope* of coalitions, we argued that groups focusing on climate-related topics should be more cohesive than more generic groups. If countries join forces specifically around shared climate-related interests, this should lead them to more often find common positions on these issues, and therefore be more important in the UNFCCC negotiations. Indeed, we see negative and statistically significant effects for generic coalitions across all four models, indicating that such coalitions are less central on all four centrality measures than climate-specific groups. This effect is also the strongest of the binary variables in three of the models (in the betweenness model, it is a close second), with the difference between climate-specific and generic coalitions representing around 20% of the total variation of those dependent variables (in the betweenness case it is 12%). Thematic scope is thus a highly important determinant of the role coalitions play in the climate negotiations, and the respective hypothesis is corroborated.

Coalition size increases the power of coalitions, as it allows members to bring together resources from more partners. The models substantiate this expectation, as all effects are positive and significant, meaning that larger coalitions are more central on all four measures of centrality. As membership varies in the dataset from only three (Mountainous Landlocked Developing Countries) to 129 (G77), this effect is the strongest overall in the models. For instance, in the in-degree model, when keeping all other variables at their means, the smallest coalitions are predicted to have a centrality score of 0.04, which increases to 0.18 for the largest coalition. This is about 45% of the total possible variation of this variable. This magnitude of the effect is similar for the other models, but it should be kept in mind that there is only one such large coalition in the dataset, while all other coalitions have below 60 members. If we limit the analysis to these smaller coalitions, coalition size still remains an important determinant for how central coalitions are in the climate negotiations, yet the effect size is no longer the largest in the models.

While we found evidence favouring the three hypotheses discussed so far, the idea that more formal coalitions play a more important role in the climate negotiations is not corroborated. We argued that, for instance, official structures (such as a secretariat) should provide members with resource-related advantages. However, none of the coefficients on the variable level of formality is close to significance, suggesting that, all else being constant, coalitions' level of formality does not affect their roles in the climate negotiations.

Finally, the coalition activity control variable is positive and significant in the first three models, indicating that older coalitions take on more central roles, while in the Closeness model the effect is insignificant and very close to zero. The reason for the latter finding might be that while the older coalitions are more central in the networks in general, the newer coalitions are connected just as well between them and with their member countries, and thus have larger closeness scores (see particularly the last network for Figure 3.1). However, for this last finding we suggest additional research should be carried out in the future.

Conclusions

This study is a first attempt to systematically identify the role that different coalition types adopt in the international climate negotiations. Our central assumptions are, first, following the ACF, that coalition types that bind the member countries more closely together (via biased assimilation and fostering cohesion) play a more central role in the negotiation, and second, according to the RDT, that coalition types that enable countries to bundle more resources and thus gain more power will also play a more central role. Based on these assumptions, and starting from the classification of coalitions as proposed by Castro and Klöck (2020), we derive expectations on various types of coalitions. We show that, as expected, geographic and thematic scope, as well as coalition size play an important role in determining how central coalitions are in the negotiations. Regional, climate-focused, larger, and older coalitions tend to play a more central role in the negotiations, both in terms of their levels of activity and popularity (out-degree and in-degree) and in terms of how they act building bridges between their members and all other parties (betweenness). In addition, regional, climate-focused, and larger coalitions seem to adopt a position that is closer to more players in the negotiations (closeness).

As stated above, this study represents just a first step towards a more systematic research agenda on coalition behaviour in international negotiations. Further research could add more depth by not only looking at average coalition-level characteristics, but also investigating whether the characteristics of their individual members are also relevant, and looking at whether, and to what extent, members of the same coalition also have an increased likelihood of working together bilaterally. The period of analysis could also extend beyond the year 2013, which is the last year available in our data. Besides this, we propose more detailed research into the circumstances under which coalitions tend to adopt common positions (e.g. with respect to what topics, or at what times in the negotiation process), and on how successful they are in achieving their goals. A key limitation of the study is that by focusing on data from the ENBs that only describes the open oral negotiations, we cannot include in the analysis more informal patterns of cooperation and negotiation interaction in which coalitions also engage. Further research on coalitions should therefore also strive to take a look at these other more informal channels of influence, which are at least as important as the official negotiations.

Notes

1 Note that because the existing dataset only covers the UNFCCC negotiations until 2013, it does not include some of the most recently created coalitions, such as the High Ambition Coalition.
2 Note that given that the climate change negotiations are based on consensus, each party actually has the power to veto an emerging agreement. Thus, while each country can prevent a negotiated deal, no country alone is powerful enough to prevail on a negotiated deal passed by the UNFCCC.
3 This view is disputed, as some researchers also claim that countries within regions do not have converging climate interests (see Richards, 2001).
4 Some ENBs during this period were not coded. These include the first ENB for each negotiation round, because it offers a summary of the previous developments in the negotiations, and the last ENB for each negotiation round, because it summarises what was already reported in the individual issues during the previous days, and in addition describes the main decisions and agreements reached during the meeting. The dataset includes meetings of the COP, but also meetings of more specific UNFCCC bodies (such as the subsidiary bodies in charge of discussing implementation-related and science-related issues, the ad-hoc working groups mandated to reach specific new agreements, or individual workshops on specific topics under discussion).
5 Note that even though coalitions, such as AOSIS or the EU, represent groups of countries putting forward a common position or statement, in this analysis we treat them as individual actors themselves. The idea behind this approach is to assess how central the coalitions themselves are in the negotiation network, which we use as a measure of the extent to which the coalitions actually fulfil their role of representing the views of their members.
6 In this analysis we disregard the far less common conflictive ties, because a different set of explanatory factors would be required to explain centrality with respect to conflictive negotiation interactions.
7 The difference between the number of members reported here, and the official number of groups members as reported in Klöck et al. (2020) reflects, on the one hand, the fact that this study counts the countries actively participating in the UNFCCC negotiations, and on the other, that group membership has changed over time (for instance, South Sudan joined the G77 after the end of the present study).
8 With this last variable we aim to capture whether coalitions are "old", i.e. involved for a long time in the climate negotiation, or relative newcomers. We simply use a single value for the overall number of active years, for instance, 19 for G77, and include this value in the regression analysis. However, even when we code this in a more complex way, i.e. adjust the value by 1 for every additional active year, the results do not change.

References

Antonides, G. (1991). Psychological variables in negotiation. *Kyklos, 44*(3), 347–362.

Ashe, J. W., Van Lierop, R., & Cherian, A. (1999). The role of the alliance of small island states (AOSIS) in the negotiation of the United Nations framework convention on climate change (UNFCCC). *Natural Resources Forum, 23*(3), 209–220.

Bailer, S. (2004). Bargaining success in the European Union: The impact of exogenous and endogenous power resources. *European Union Politics, 5*(1), 99–123.

Betzold, C. (2010). 'Borrowing' power to influence international negotiations: AOSIS in the climate change regime, 1990–1997. *Politics, 30*(3), 131–148.

Betzold, C., Castro, P., & Weiler, F. (2012). AOSIS in the UNFCCC negotiations: From unity to fragmentation? *Climate Policy, 12*(5), 591–613.

Blaxekjær, L. Ø. (2020). Diplomatic Learning and Trust: How the Cartagena Dialogue Brought UN Climate Negotiations Back on Track. In C. Klöck, P. Castro, & F. Weiler (Eds.), *Coalitions in the Climate Change Negotiations*. Abingdon: Routledge.

Castro, P. (2017). Relational data between parties to the UN framework convention on climate change.

Castro, P., & Klöck, C. (2020). Fragmentation in the Climate Change Negotiations: Taking Stock of the Evolving Coalition Dynamics. In C. Klöck, P. Castro, F. Weiler, & L. Ø. Blaxekjær (Eds.), *Coalitions in the Climate Change Negotiations*. Abingdon: Routledge.

Ciplet, D., Roberts, J. T., & Khan, M. R. (2015). *Power in a Warming World: The New Global Politics of Climate Change and the Remaking of Environmental Inequality*. Cambridge, MA and London: MIT Press.

Clauset, A., Newman, M. E., & Moore, C. (2004). Finding community structure in very large networks. *Physical Review. part E, 70*(6), 066111.

Csardi, G., & Nepusz, T. (2006). The igraph software package for complex network research. *InterJournal, Complex Systems*, 1695(5), 1–9.

DeSombre, E. R. (2000). Developing country influence in global environmental negotiations. *Environmental Politics, 9*(3), 23–42.

Dillenbourg, P., Baker, M. J., Blaye, A., & O'Malley, C. (1995). The evolution of research on collaborative learning. Retrieved from https://telearn.archives-ouvertes.fr/hal-0019 0626/

Drahos, P. (2003). When the weak bargain with the strong: Negotiations in the World Trade Organization. *International Negotiation, 8*(1), 79–109.

Dupont, C. (1994). Coalition Theory: Using Power to Build Cooperation. In W. Zartman (Ed.), *International Multilateral Negotiations: Approaches to the Management of Complexity* (pp. 148–177). San Francisco: Jossey-Bass Publishers.

Dupont, C. (1996). Negotiation as coalition building. *International Negotiation, 1*(1), 47–64.

Freeman, L. C. (1979). Centrality in social networks conceptual clarification. *Social Networks, 1*(3), 215–239.

Goodreau, S. M., Kitts, J. A., & Morris, M. (2009). Birds of a feather, or friend of a friend? Using exponential random graph models to investigate adolescent social networks. *Demography, 46*(1), 103–125.

Hafner-Burton, E. M., Kahler, M., & Montgomery, A. H. (2009). Network analysis for international relations. *International Organization, 63*(3), 559–592.

Hamilton, C., & Whalley, J. (1989). Coalitions in the Uruguay round. *Weltwirtschaftliches Archiv, 128*(3), 547–562.

Henry, A. D. (2011). Ideology, power, and the structure of policy networks. *Policy Studies Journal, 39*(3), 361–383.

Hopmann, P. T. (1996). *The Negotiation Process and the Resolution of International Conflicts.* Columbia: University of South Carolina Press.

IISD. (1995–2013). Earth negotiations bulletin. Retrieved from http://enb.iisd.org/enb/vol12/

Klöck, C., Weiler, F., & Castro, P. (2020). Appendix 2: Coalitions in the Climate Negotiations. In P. Castro, C. Klöck, L. Ø. Blaxekjær, & F. Weiler (Eds.), *Coalitions in the Climate Change Negotiations.* Abingdon: Routledge.

Krasner, S. D. (1991). Global communications and national power: Life on the Pareto frontier. *World Politics, 43*(3), 336–366.

Milner, H. (1992). International theories of cooperation among nations: Strengths and weaknesses. *World Politics, 44*(3), 466–496.

Muldoon, J. (2018). *Multilateral Diplomacy and the United Nations Today.* Boulder: Routledge.

Narlikar, A. (2004). *International Trade and Developing Countries: Bargaining Coalitions in GATT and WTO.* London, New York: Routledge.

Olson, M. (1965). *The Logic of Collective Action.* Cambridge, London: Harvard University Press.

Ott, H. E., Bauer, S., Brandi, C., Mersmann, F., & Weischer, L. (2016). Climate alliances après Paris: The potential of pioneer climate alliances to contribute to stronger mitigation and transformation. Retrieved from https://hermann-e-ott.de/cms/wp-content/uploads/2016/11/PACA_paper_2016_WI-1.pdf

Panke, D. (2013). *Unequal Actors in Equalising Institutions: Negotiations in the United Nations General Assembly.* Basingstoke: Palgrave Macmillan.

Richards, M. (2001). *A Review of the Effectiveness of Developing Country Participation in the Climate Change Convention Negotiations.* London: Overseas Development Institute.

Rosenzweig, C., Karoly, D., Vicarelli, M., Neofotis, P., Wu, Q., Casassa, G., . . . Seguin, B. (2008). Attributing physical and biological impacts to anthropogenic climate change. *Nature, 453*(7193), 353.

Rubin, J. Z., & Zartman, I. W. (2000). *Power and Negotiation.* Ann Arbor, MI: University of Michigan Press.

Sabatier, P. A., & Jenkins-Smith, H. C. (1993). *Policy Change and Learning: An Advocacy Coalition Approach.* Boulder, CO: Westview Press.

Snyder, G. H., & Diesing, P. (2015). *Conflict Among Nations: Bargaining, Decision Making, and System Structure in International Crises.* Princeton: Princeton University Press.

Tobin, P., Schmidt, N. M., Tosun, J., & Burns, C. (2018). Mapping states' Paris climate pledges: Analysing targets and groups at COP 21. *Global Environmental Change, 48,* 11–21.

Valente, T. W., Coronges, K., Lakon, C., & Costenbader, E. (2008). How correlated are network centrality measures? *Connections, 28*(1), 16–26.

Venturini, T., Baya Laffite, N., Cointet, J.-P., Gray, I., Zabban, V., & De Pryck, K. (2014). Three maps and three misunderstandings: A digital mapping of climate diplomacy. *Big Data & Society, 1*(2), 2053951714543804.

Wasserman, S., & Faust, K. (1994). *Social Network Analysis: Methods and Applications.* Cambridge: Cambridge University Press.

Watts, J. (2020). AILAC and ALBA: Differing Visions of Latin America in Climate Change Negotiations. In P. Castro, C. Klöck, L. Ø. Blaxekjær, & F. Weiler (Eds.), *Coalitions in the Climate Change Negotiations*. Abingdon: Routledge.

Weible, C. M. (2005). Beliefs and perceived influence in a natural resource conflict: An advocacy coalition approach to policy networks. *Political Research Quarterly*, *58*(3), 461–475.

Weiler, F. (2012). Determinants of bargaining success in the climate change negotiations. *Climate Policy*, *12*(5), 552–574.

Zartman, I. W. (1994). Introduction: Two's Company and More's a Crowd: The Complexities of Multilateral Negotiation. In W. Zartman (Ed.), *International Multilateral Negotiations: Approaches to the Management of Complexity* (pp. 1–12). San Francisco: Jossey-Bass Publishers.

Zartman, I. W. (1997). The Structuralist Dilemma in Negotiation. In R. J. Lewicki, R. J. Bies, & B. H. Sheppard (Eds.), *Research on Negotiation in Organizations* (Vol. 6, pp. 227–245). Greenwich: JAI Press.

4 The temporal emergence of developing country coalitions

Nicholas Chan

Introduction

Over more than a quarter-century of international climate change negotiations, the North–South divide between developed and developing countries has been a central part of the political dynamics of the climate negotiating process (Roberts & Parks, 2007). This divide has been routinely described through terms such as "mistrust" and "polarisation", especially highlighting the issue of equity and whether the outcomes from the climate negotiating process were "fair" and "equitable" for developing countries (Chan, 2016a). But a key feature of how this divide has manifested in the United Nations Framework Convention on Climate Change (UNFCCC) negotiating process has been through the pattern of negotiating coalitions of countries that, for the most part, have remained divided along North–South lines: the Group of 77 and China (G77) coalition has remained the umbrella negotiating group for developing countries, while new coalitions that have formed have remained among developing countries, with substantial overlaps, rather than across North–South lines. What explains this pattern?

A conventional rationalist interpretation suggests that the determinants of a state's negotiating position, including on environmental issues, are a weighing up of its material costs and benefits. In an environmental context, it is the relative balance between mitigation costs and ecological vulnerability to adverse impacts that shapes positions towards strong or weak international agreements on environmental regulation (Buys, Deichmann, Meisner, That, & Wheeler, 2009; Sprinz & Vaahtoranta, 1994). States that share similar cost/benefit situations might therefore be expected to align together to express common negotiating positions through forming coalitions and alliances. Developing countries have faced diverse cost/benefit situations, with uneven mitigation costs and ecological vulnerabilities to climate change across the more than 130 countries of the Global South. This heterogeneity of interests among developing countries has long been observed, and is not itself a new phenomenon (Williams, 2005).

This chapter, however, argues that the way in which these cost/benefit preferences and divergences are expressed through coalition-building is also shaped by the institutional context of negotiations themselves: *where* negotiations are conducted, *how* they are organised, how the legacy effects of the institution shape choices over *what kind* of coalition to form, and *when* to do so. What states want out of international climate negotiations is also shaped by the structure and context of the negotiating process itself. While "process" is often regarded as an essential element of explaining negotiation outcomes (Monheim, 2014), its implications for coalition activity among developing countries have been less well-discussed.

This chapter draws on the growing interest in historical institutionalism (HI) within international relations (Fioretos, 2011) to trace how the institutional context shapes the pattern and timing of coalition emergence in the climate negotiations. HI has sought to explain particular features in the emergence and evolution of questions of international order, or of international organizations themselves (Farrell & Finnemore, 2017, pp. 392–394; Fioretos, 2011). There has been some examination of specific issues within the climate negotiations through a HI perspective, such as on deforestation (Pistorius, Reinecke, & Carrapatso, 2017), and there is a brief recognition of path-dependent dynamics in the G77 context by Vihma, Mulugetta, and Karlsson-Vinkhuyzen (2011, pp. 318, 332). In general, however, this has been an under-utilised approach to analysing the politics of the climate negotiations. This chapter also goes further theoretically by treating coalitions and alliances themselves as institutions, rather than the tendency in major HI work to focus on formal international *organisations* (Farrell & Finnemore, 2017). This approach therefore aims to provide a more micro-level application of HI dynamics where the characteristics that have been at the centre of macro-institutional analysis, such as political space, set-up costs, and layering, are also relevant in understanding interactions between states *within* a particular institutional context. To illustrate this briefly, the G77 itself is recognised as an institution, rather than just a reflection of objective shared interests – it does not have an especially thick organisational form, but is nonetheless an "informal yet highly institutionalised mechanism" (Williams, 1997, p. 296). Despite the emphasis placed by some interpretations of coalitions as being temporary and fleeting (Chasek, 2005, p. 126), they are also sites that fit the broad understanding of institutions as bundles of norms, principles, and rules that reflect shared understandings and expectations.

The particular contribution and argument of this chapter is to highlight the *temporal* aspect of coalition dynamics – temporality, put simply, is that "the timing and sequence of events shape political processes" – being one of the key emphases of HI (Fioretos, 2011, p. 371). Choices and moments of formation of new coalitions are not just about the alignment of material interests, although these are obviously significant. Explaining the *emergence* of developing country coalitions is also a question of temporality – why do particular coalitions form at particular points in time? How do existing coalitions shape subsequent coalition formation? In particular, why have coalitions taken an overlapping format, being "layered" on top of each other?

This chapter addresses these questions by tracing two dynamics that show how the institutional context of the UNFCCC process has affected the temporal pattern of the emergence of developing country coalitions. The first section of this chapter returns to the starting point of the climate negotiations themselves, as the institutional founding moment that defined the landscape of possible coalition patterns.

This section highlights the role of the G77 as the "default" coalition, given the institutional mandate for negotiations within the UN General Assembly (UNGA), which dominated the "political space" available for subsequent coalition formation, as well as contributing to the political construction of climate change as a North–South issue. This dominance of political space has meant that efforts to form new coalitions have only been possible through the treatment of these coalitions as "subgroups" of the G77 that form a second – or even third – layer of coalition membership and participation, on top of the foundational layer of the G77. The second section then turns to the process of negotiation itself, and how institutional rules for organising negotiations also shape the coalitions that form, through an emphasis on coalition-based negotiation. The negotiating process provides incentives for coalition formation, particularly in a cascading, "reactive" dynamic. This is particularly demonstrated in post-2009 coalition formation, for states that did not have a second "layer" of membership to then seek to form a coalition. By examining these two broader dynamics of the negotiating process as a whole, this chapter also aims to complement the attention paid to specific coalitions in other chapters of this volume.

The emergence of developing country coalitions: founding moments and the centrality of the Group of 77

A recurring theme in the academic literature on the UNFCCC process is the question of why developing countries have been able to "hang together", despite their diversity (Barnett, 2008; Hochstetler, 2012, p. 59). Focusing on how negotiations were organised from the outset in 1991 provides one way of addressing this question, which is more than just historical curiosity. It points to how the G77 was the default mode of organising developing countries, which is essential to understanding its subsequent durability as the umbrella grouping for developing countries.

A core concept in HI approaches is sequencing: that the order in which political events happen has a causal effect upon later developments (Fioretos, 2011, p. 381). Initial decisions have implications in either widening or narrowing the menu of available choices for subsequent decisions. While path dependence in terms of self-reinforcing positive feedback from a particular decision point has been the common emphasis in much HI scholarship, negative feedback or "reactive sequences" that *undermine* the initial decision may also be the dynamic at play (Hanrieder & Zürn, 2017). This attention to initial decisions points to the need to look back to the starting point of the negotiating process itself, and what the pattern of coalitions has evolved *from*. Coalition choices, from this perspective, are relational, depending on what existed beforehand, rather than constructed on just a snapshot interpretation of material preferences. As Pierson (2004, pp. 71–74) suggests, coming first to a political process can provide advantages to "first mover" actors in filling up "some limited 'political space'", making it more difficult for other actors – in this case coalitions – to displace initial ones.

The first part of this section shows how this was the case in terms of the G77 as the first-mover coalition that encompassed *all* developing countries. However, it also shows how some of the G77's own institutional rules for coordinating also allowed subgroups to subsequently form. The second part of this section then points to the pattern demonstrated by these subsequent coalitions that follows from the G77's first-mover position: the "layering" of new developing country coalitions on top of the G77. This draws upon another HI concept to explain gradual institutional change, where "new rules are attached to existing ones, thereby changing the ways in which the original rules structure behaviour" (Mahoney & Thelen, 2009, p. 16). In the context of coalition behaviour, layering implies that membership of newer coalitions does not terminate G77 association. Instead, it resulted in a general pattern of newer intra-South coalitions, rather than coalitions across the North–South divide.

The first climate negotiations and the Group of 77

The first formal climate change negotiations were organised through a 1990 UNGA mandate in Resolution 45/212 that established an Intergovernmental Negotiating Committee (INC), setting as deadline the 1992 Rio Earth Summit (United Nations, 1990). This was not a foregone choice: during informal negotiations as climate change rose up the political agenda in the late 1980s, a variety of different processes were considered for how to organise the negotiations. A prime candidate had been the UN Environment Programme (which had successfully convened the Montreal Protocol negotiations on ozone depletion), and another one was the newly established Intergovernmental Panel on Climate Change (Chan, 2013, pp. 90–92). The relevant implication of the mandate for negotiations being established through the UNGA was that the G77 became the default coalition structure through which developing countries were organised from the very beginning of the process. This was consistent with the broader institutional norms and practices of how developing countries organised themselves within UNGA processes along North–South lines (Weiss, 2009), and thus the G77 chairmanship in the climate negotiations was assumed by the same country chairing the G77 in the UNGA. Pierson (2004, p. 73) argues that "start-up costs are often very high in collective action processes, and thus constitute a major barrier to entry"; the argument here is that the G77 provided a ready-made coalition structure for developing countries with minimal – perhaps zero – start-up costs that was also socially consistent with other patterns of cooperation within the same institution. In this way, the "choice" of the G77 may not even have really been an active choice, but more of an unreflective act along the lines of Hopf's (2010) "logic of habit".

Substantively, the choice of G77 as the coalition also had implications for the INC process, especially in shaping the political understanding of climate change as a North–South problem. Positions advocating for "common but differentiated responsibilities and respective capabilities" (CBDR-RC) and the emphasis on Northern obligations to provide "means of implementation" for developing countries (financial support, technology transfer, and capacity-building support) drew on the generalised concern coming out of North–South conflict in the 1980s about preserving the "development space" of the South amid the emergence of the "sustainable development" norm (Najam, 1994; Ramakrishna, 1992). Marc Williams' argument about the "re-articulation" of the South is especially important (Williams, 1993); in referencing the broader efforts of developing countries to caucus together within the UN system since the 1960s when the G77 was established, Williams also points to the broader social identity of marginalisation in international affairs that the G77 aimed to express, leading towards positions in the climate context that also aimed to redress broader economic marginalisation and safeguard development prospects. Additionally, the Montreal Protocol negotiations, seen as a model for the climate negotiations, also provided lessons about the possibilities of successful Southern bargaining over the global atmospheric problems (Hoffmann, 2005), further motivating developing countries to seek a united approach within a G77 structure (and also further illustrating the importance of relative timing and sequencing).

The development of these positions through a G77 structure therefore also introduced a degree of rigidity into the political choices over coalitions that developing countries had: participation in the G77 in the INC negotiations was automatic, given wider G77 membership in the UNGA. By way of illustration, at the first INC meeting in February 1991, just 66 developing countries participated – barely half the G77 membership – and yet they caucused and spoke under the G77 banner (Chan, 2013, p. 127). Any "latecomer" developing countries to the INC process would have been expected to coordinate within the G77 and support the past positions agreed by the group up until that point. In effect, the G77 dominated and occupied the entire possible political space for developing country coalitions, where no opt-out was possible.

Interpreting the G77 as "default" coalition for developing countries is also reflected in two further features of behaviour. The first is that even after differences emerged at particular negotiating sessions, as the agenda moved on to a different question, the G77 continued to convene. For instance, one of the most well-known incidents of a divided G77 was at COP1 in 1995 over the scope of the "Berlin Mandate" for what became the Kyoto Protocol, which saw disagreement among G77 members, especially island states (who had grouped together as the Alliance of Small Island States (AOSIS) and oil-exporting countries. India, the Philippines, and a smaller subset of developing countries then formed a narrower "Green Group" (also including AOSIS) that forged a compromise agreement, excluding the oil-exporting countries (Sengupta, 2012). But once that compromise led the way to an overall agreement by the conference as a whole, and the question of the mandate was essentially settled, the G77 returned to caucusing over the new questions that the Berlin Mandate working group faced. Similarly, the link between adaptation and "response measures" (the negative impacts of climate mitigation actions) during the Kyoto Protocol negotiations was the source of considerable division, resulting in only minimal G77 positions (Mwandosya, 1996). Continued differences among subgroups with divergent positions did not result in a terminal break for the G77 on this question, which was subsequently able to find (uneasy) compromises (Barnett, 2008; Chan, 2016b).

Conversely, while the differences among G77 members have often been emphasised in tracing the formation of new coalitions, the continuing existence of the G77 as the default coalition has provided a focal point through which to explore potential areas of *greater* convergence in these positions as the negotiating agenda has evolved. For instance, the emergence of loss and damage in the pre-Paris Agreement negotiations had been historically championed by AOSIS and the Least Developed Countries (LDCs), but also became an area of G77 coordination and common positions (Fry, 2016). Finance has been the other area where a cohesive substantive position has been agreed among the G77. Despite differences among developing countries over definitions of vulnerability and prioritising particular subgroups of developing countries as being "particularly vulnerable" for allocating adaptation finance (Betzold & Weiler, 2018), developing countries have maintained common positions on the evolution of new climate finance institutions and procedures, especially vis-à-vis developed countries.

Beyond particular issues, this automatic, default nature of the G77's establishment from the outset of the INC negotiations is also illustrated by comparison with the one other developing country group that also formed at that initial moment: AOSIS. Pierson's observation about the start-up costs of collective action cited above (Pierson, 2004) contrasts with the considerable political entrepreneurship required to mobilise this entirely new group of countries in the UN system. Most accounts of AOSIS rightly point to the shared vulnerability among island states, particularly in terms of sea-level rise, as justification for AOSIS' formation (Ashe, van Lierop, & Cherian, 1999; Betzold, 2010). But this emphasis on the material underpinnings of AOSIS also need to be appreciated for the agency that was required to bring together countries across three regions (the Caribbean, Pacific, and Mediterranean/Indian Oceans) that did not have much history of prior coordination and had some quite different economic circumstances, at a time before the *small island developing states (SIDS)* category was institutionalised within the UN system (Chasek, 2005). It was at one of the pre-INC informal meetings, the Second World Climate Conference in Geneva in 1990, when Caribbean countries met with delegates of other small island states, through the initiative of Trinidadian environment minister Lincoln Myers, and issued a ministerial declaration as "AOSIS" for the first time at the close of this conference (Chan, 2013, p. 112). This pre-INC coordination among small island states therefore illustrates the political effort that had to go into overcoming political inertia and establishing a new group, rather than having an existing structure to rely on.

Reflecting the limited political space available, however, AOSIS was also careful to not diverge too far from the G77: coordinating within AOSIS was not a substitute for coordinating with the G77 because to do so would have disavowed broader G77 membership. The choice of Vanuatu as first AOSIS chair was instructive of this balance to be struck between the agency of the new coalition, and default membership of the overarching group (Chan, 2013, p. 130–131). Vanuatu and its Permanent Representative to the UN, Robert van Lierop, had been a prominent G77 voice in the UNGA's decolonisation debates, and its chairmanship was intended to assuage wider concerns among the rest of the G77 membership that

AOSIS was a more far-reaching split from the G77 itself. Instead, from this beginning, AOSIS sought to make the case that it was a *subset* of the G77, rather than a splinter from it.

"Layering" and the possibility of subgroup coalitions

This initial pattern of coalition formation at the very beginning of the INC negotiations also demonstrates the dynamics surrounding timing and subsequent coalition formation among developing countries: "layering" new coalitions on top of the G77, such that coordination within these newer groups was not exclusive of countries' existing G77 membership within the wider UNGA process and UN system. It also, however, reflects a consequence of this "default" character to the G77: that a country cannot opt out of the G77's agreed positions in isolation, but it can add a layer to it.

This possibility of layering also reflected the G77's own institutional practices, which allowed some flexibility for its members, principally the use of regional subgroups as a method of caucusing (Williams, 1997). This allowed the possibility of subsets of the G77 to form and present distinct positions without disassociating themselves from the larger group itself that still presented common positions, where these exist, on the same issue. This practice of treating newer coalitions as subgroups *of the G77* rather than distinct groups in their own right is reflected in how these smaller coalitions participate within the negotiations themselves: interventions by any of these subgroups make a point of "associating" themselves with the prior intervention delivered by a G77 spokesperson or coordinator, who by custom delivers the first intervention. This is common practice to the extent that if the G77 spokesperson is delayed in arriving into the room, spokespersons of subgroups will generally refrain from delivering their own interventions, even if that can mean delaying overall proceedings (personal observation, 2012; Yamin & Depledge, 2004, p. 36). Indeed, treating other coalitions as subgroups also means that they can intervene *during* G77 coordination meetings to speak on behalf of their subgroup in attempting to shape the overall G77 position (personal observation, 2012). Treating these other coalitions as subgroups rather than as distinct, splinter coalitions has an important implication for the broader pattern of coalition formation: if a coalition was a subgroup of the G77, it could not include countries that were *not* members of the G77. This helps to account for layering outcomes such that that all standing, formal subgroups that have been formed by G77 countries have been almost exclusively among G77 members (with the exception of four Pacific small island developing states who are nonetheless AOSIS members: Niue and the Cook Islands, who are not UN member states, and Palau and Tuvalu).

While AOSIS was the only formal developing country coalition other than the G77 at the beginning of negotiations, these institutional practices smoothed the path for the emergence of subsequent coalitions, especially those based around pre-existing subgroups, to join AOSIS on this "second" layer of coalition participation on "top" of the G77. The establishment of the LDC group in 1999, as negotiations expanded to discuss LDC-specific issues within the UNFCCC process (such as the LDC Fund, the National Adaptation Programmes of Action process, and the LDC Expert Group), did not come at the expense of G77 membership (Yamin & Depledge, 2004, p. 36). Instead, it reflected the well-established institutional category of LDCs within and beyond the UN system as a subgroup of developing countries (Fialho, 2012), as well as the special treatment explicitly for LDCs already recognised in Article 4.9 of the Convention. Similarly, while the African countries also coordinated through the African Group of Negotiators (AGN) structure, and became particularly active from the mid-2000s onwards (Roger & Belliethathan, 2016), their formation was relatively uncontroversial, because it was based on the existing African regional subgroup within the G77 itself. Both these relatively large subgroups – the LDCs with 47 members, Africa with 54 members – overlapped even further, with most LDCs being African countries. As a result, some countries have found themselves simultaneously as members of three groups, one layered on top of the other: the LDCs, AGN, and G77. This type of situation, and the other possible variants of multiple coalition membership, demonstrates the flexibility (although separate from questions of coalition efficiency) that was possible in the G77's institutional practices, resulting in layered coalitions.

The importance of the G77's temporal advantage in "coming first" as the underlying, foundational coalition, and its role in shaping the North–South divide itself, can also be contrasted with the case of the one longstanding coalition comprising both developed and developing countries: the Environmental Integrity Group (EIG). The two developing countries in this coalition, Mexico and the Republic of Korea, who were also not included in Annex I of the UNFCCC, also ceased to be G77 members after joining the Organisation for Economic Cooperation and Development (OECD) in the mid-1990s. Being outside the G77 therefore meant that they were not subject to these practices and associations with the G77, but neither could they participate in G77 subgroups. Their existence reflects their position amid these institutional categories: one description characterises the EIG as "strange bedfellows, sharing little in terms of national circumstances except for the fact that they *do not belong to any of the other main groups*" (Yamin & Depledge, 2004, p. 48, emphasis added). As will be highlighted in the next section, being part of a group is a necessity due to how the negotiating process itself is organised, further showing how institutional practices provide incentives for coalition formation.

A potentially more disruptive moment for this layering dynamic, however, was the emergence later into the 2000s of newer groups that did not reflect existing subgroups, categories, or structures for coordination, particularly Brazil, South Africa, India, and China (BASIC) at the 2009 Copenhagen conference. The material reasons for BASIC's emergence as a distinct group have been explored elsewhere (Hallding, Jürisoo, Carson, & Atteridge, 2013), but BASIC was also important in this temporal sequence of coalition formation: it opened up a greater possibility for G77 subgroup coalitions that were *not* based around existing categories or structures. One of BASIC's immediate challenges, especially in the aftermath of Copenhagen and amid cooperation with the USA in shaping the Copenhagen Accord, was its relationship with the rest of the G77. Other developing countries had objected to the adoption of the Accord itself at Copenhagen (where it was instead only "taken note" of). The broader "emerging economy" status that was beginning to gain traction in world politics also posed the question whether BASIC countries were engaging in a "graduation" of sorts from the G77 (Hochstetler, 2012). BASIC's response to this challenge was to then present itself as another subgroup of the G77, in similar terms to the then-existing AOSIS, LDCs, AGN, and ALBA (Bolivarian Alliance for the Peoples of Our America) groups. BASIC quarterly ministerial meetings included participation from other such G77 subgroups as well as the G77 chair in a formulation termed the "BASIC Plus" approach, which began at its fourth ministerial meeting in 2010 (Brazil, 2010). At this meeting, continued G77 association was reiterated in a BASIC joint statement, in which ministers "highlighted the role of BASIC as part of the Group of 77 and China and the importance that the BASIC countries maintain their full participation in the discussions and activities of the Group" (Brazil, 2010). This conscious effort to retain its G77 association has its instrumental dimensions for BASIC itself, as many other developing countries would also want to retain the bargaining clout of these countries on their side (Hochstetler, 2012, pp. 60–62). The ability of the BASIC group to fit itself in this way into the existing structure of the G77, however, reiterates the argument of this chapter about the importance of the institutional practices that shape and enable coalition formation. This moment also set in process a greater permissiveness for subgroups to form, as was subsequently observed with the cases of some other groups further discussed in the next section, that were not based around existing UN classifications (such as the LDCs) or regional categories (such as the AGN).

In this regard, the salient feature to note about the final year of Paris Agreement negotiations, and the role of the "High Ambition Coalition" (HAC) that has been recognised as playing a particularly important role, is that there was a need to maintain "secrecy" for developed and developing countries when more closely coordinating their positions. For instance, private informal dinners convened by Marshall Islands minister Tony de Brum (Mathiesen & Harvey, 2015) have been highlighted as being pivotal in cementing coordination among its members at COP21. But this secrecy also points to the continuing relevance of G77 exclusivity – that to consistently and publicly coordinate positions with developed countries would violate established practices about only establishing formal coalitions comprised of other G77 members. The looser, less formal structure of the HAC, where it has avoided making plenary statements, is in part a reflection of continued default membership of the G77 for its developing country participants. A precursor to the HAC, the Cartagena Dialogue for Progressive Action, also had to navigate these practices surrounding G77 membership, and for the most part similarly eschewed making plenary statements and sought to avoid being portrayed as a formal negotiating coalition (Blaxekjær, 2020).

In short, the sequence of coalition formation has been temporally marked by two important moments: the establishment of the G77 as the "default" coalition within which to organise developing countries, but including the flexibility to manage subsequent differences and divergences in its recognition of layered subgroups; and the establishment of BASIC, and the greater permissiveness afforded to forming G77 subgroups beyond existing categories. The North–South divide is not one that has taken place within an institutional vacuum, but where the institutional context for negotiation itself contributed to constituting this divide.

Organising negotiations through coalitions in the UNFCCC process

The longer-term consequence of continued layering of new coalitions has interacted with other dynamics of the process, namely, how negotiations themselves are organised, that gives priority to coalitions. This can be understood as a "reactive" sequence, as introduced above, whereby elements of the original institutional choice can also, over time, undermine the initial all-encompassing coalition choice – in this case, the G77. The result is that momentum to form new coalitions becomes self-reinforcing: as newer coalitions were formed, a coalition "cascade" followed as a result of the institutional need for second-layer representation within the negotiating process. This shows the significance of sequencing: as more developing countries began to express their positions through a second layer of narrower, subgroup coalitions in addition to the G77, this had negative implications for those countries that did *not* participate in a grouping within this second layer.

The organization of negotiations

The UNFCCC is often highlighted as a prime example of "universal" multilateralism, with participation by 197 parties. Together with its consensus decision-making process (in the absence of the formal adoption of rules of procedure), this is hailed as the basis of its inclusivity, allowing for fair and democratic participation and decision-making: for an issue that affects everyone on the planet, everyone has a say. At the same time, both of these features have been the subject of criticism, especially after the collapse of the 2009 Copenhagen conference, surrounding the inability of such a decision-making structure to result in agreement. Various alternative "minilateral" configurations have been proposed as a way of efficiently reaching agreement (see especially Eckersley, 2012).

But what is often overlooked in this discussion, or at least in the general portrayal of the climate negotiations as a negotiation among 197 parties, is that in practice the process is already a heavily minilateral one through the organisation of coalition participation to aggregate negotiating positions. For instance, groups of countries are privileged over individual states in the order in which plenary statements are delivered. Over the course of a negotiating conference, building small groups – often organised through coalitions – has always been the method to actually reach agreement (Bodansky, Brunnée, & Rajamani, 2017, pp. 75–82). Contentiousness has arisen over who is included in these small groups, over different criteria of representativeness (Eckersley, 2012, p. 37). This in turn generates certain incentives for coalition formation and participation: "The demand by countries to form new coalitions also responds to the growing tendency of structuring negotiations based on coalitions" (Yamin & Depledge, 2004, p. 34).

The typical sequence of negotiation runs from open plenaries to contact groups and "informal" settings (which have taken various names and configurations over the years) overseen by facilitators, chairpersons of subsidiary bodies, or the conference presidency (Bodansky et al., 2017, p. 79; Yamin & Depledge, 2004, p. 442). These facilitators commonly prepare draft text for agreement based upon views expressed, open to all parties who are interested – and after inviting further views, typically prepare second or third iterations that aim to accommodate these reactions. But on significant, contentious issues, the conduct of negotiations through informal meetings open to all is also often insufficient to reaching consensus. The next stage, typically in the final few days of negotiating conferences, often then sees facilitators conduct bilaterals with groups (Bodansky et al., 2017, p. 80). These are individual, private meetings between facilitator and individual groups to test potential solutions, and identify genuine "red lines" and potential "landing zones" in order to draft a compromise to present to a larger setting. Dimitrov (2016), for instance, notes that the final two days of the 2015 Paris Conference were conducted through this mode of work, with no formal meetings convened until the presentation of a compromise proposal by the French presidency.

Reaching actual agreement therefore depends heavily on the "penholding" power of the chair: first, the chairpersons of constituted bodies and their appointed facilitators, and second, the conference president to whom they report. Other studies have similarly drawn attention to this importance of the conference president in shaping the possibilities for agreement (and possible reforms) (Vihma, 2015). The presidency becomes particularly important in judging what the possible compromises are in order to present a "package" text as a take-it-or-leave-it situation for adoption where consensus is indicated through the absence of objections (Vihma, 2015). On occasion, the presidency has gavelled adoption of a decision through despite opposition from a single party, such as the Russian Federation at COP18 in Doha, or Bolivia at COP16 in Cancún, with the latter conference also resulting in the interpretation that "consensus" does not mean unanimity (Bodansky et al., 2017).

As a result, who the presidency consults takes on particular importance. Negotiations proceed on the mantra that "nothing is agreed until everything is agreed", providing a final opportunity to object to adoption before the president's gavel is brought down. But the more basic political fact is that no country wants to be put in the spotlight and to be seen to hold up progress, especially when the conference is typically in extra time, running over its scheduled closing (Yamin & Depledge, 2004, p. 461). The conference presidency's text becomes the default negotiating proposition (Dimitrov, 2016, p. 6), creating an imperative to be included *prior* to the presentation of this final-hour compromise proposal, and to refrain from messily objecting and attempting to block overall agreement to include particular viewpoints at a final moment (which may or may not be successful) (Vihma, 2015, p. 63).

The decision on who is invited to participate in bilaterals, in turn, is also typically based on groups that have voiced interest during the preceding sequence of negotiation. Facilitators seek out those who have expressed positions at divergent ends of the spectrum, in order to ensure their buy-in to potential compromises. Inclusion in this final phase of bilaterals depends on having expressed distinct positions during informals; group consultation during the bilateral phase depends on group participation in informals. Time constraints over the final days of negotiating conferences also mean that priority is typically given to organised groups, made possible, as an account of late-1990s negotiations observed, by the fact that: "almost all Parties to the Convention are now part of one kind of coalition or another" (Yamin & Depledge, 2004, p. 456). That said, individual countries that have strong stakes, such as the USA because of its sheer "weight" for overall outcomes, will also typically be consulted. For instance, the critical provisions in the Paris Agreement on loss and damage, and the balance between including the issue at all and potential "liability" concerns (on the part of the USA) was essentially drafted between the USA, SIDS, and LDCs (Fry, 2016) as those actors with the most forcefully articulated positions on loss and damage, before being brought back to the presidency for approval.

The structure of this process has meant that if the position of a group of countries is only minimally represented within a larger coalition, it will not have the opportunity to express the full extent of its position in this final stage *unless* it speaks out during the informal stage of negotiations. In the developing country context, the implication of this was that while the G77 would tend to have a common position, this would – because of the nature of compromise within the G77 – be the minimal, least common denominator position. Other subgroups could then go beyond this position to provide a more specific position (without disavowing the larger G77 position, as noted above).

AILAC and the LMDCs: filling in the gaps

This institutional incentive for representation is visible through the sequence of coalition formation in the 2010s that were also consequences of the *previous* emergence of other developing country coalitions In the cases of the Like-Minded Developing Countries (LMDCs, see Blaxekjær, Lahn, Nielsen, Green-Weiskel, & Fang, 2020) and the Independent Association of Latin America and the Caribbean (AILAC, see Watts, 2020), as "latecomer" coalitions, the countries that make up both groups had found themselves struggling to effectively articulate their positions within the negotiating room and bilaterals, if they only had G77 membership to draw upon for representation and input. The countries that make up both of these groups were already informally coordinating within the G77 prior to their formal group formation in order to shape the G77 position, but would be reliant on the G77 spokesperson to express their compromise G77 position. The inevitable nature of the G77 consultations and the process of developing group positions also meant that those countries that had membership of other subgroups – AOSIS, the LDCs, the AGN, and so on – in effect would have their minimal views expressed initially through the G77, but then again more directly on their subgroup basis – what might be seen as a form of double-dipping. Similarly, invitations to consult with chairs and COP presidencies would be more easily facilitated through formal declaration and mutual recognition as a group, reducing uncertainty about who was in the group that a country was speaking on behalf of. One account of AILAC, for instance, observes this motivation for AILAC's formation:

> Other countries which are part of subgroups … had the advantage of being consulted both at the general G77 meeting and at individual meetings where they were able to raise their concerns, including those where their position differed from the general G77 one. AILAC countries noted their positions were not being fully addressed under this arrangement.
>
> (Edwards, Adarve, Bustos, & Roberts, 2017, p. 74)

A similar dynamic applies in the case of the LMDCs, which was formally established concurrently with AILAC in 2012 – thus "filling in the gaps" and establishing a second layer of representation, for a swathe of developing countries, especially from Asia, that did not have prior membership of groups other than the G77. Both AILAC and the LMDCs have distinct sets of substantive positions, especially over how to interpret the CBDR-RC principle during the Paris Agreement negotiations (Edwards et al., 2017; Third World Network, 2012). For LMDC members that did have membership of other subgroups, such as BASIC (China and India), or the AGN (i.e., Sudan and Mali), the LMDCs provided a way of advancing substantive positions in the Paris Agreement negotiations that did not find consensus agreement both among the G77 as a whole, or within these other overlapping subgroups. This is, again, an institutional incentive – that their positions would have otherwise struggled to be formally expressed within the negotiating rooms without a distinct coalition (personal interview with former LMDC negotiator, 2018). Indeed, both AILAC and the LMDCs can even be seen as reactions to one another, and part of this reactive sequence. With some of their positions on differentiation being in direct opposition to each other, *not* forming a distinct coalition and being left with only G77 representation would have further diluted their ability to advance their positions. Nonetheless, in line with the argument of the first half of this chapter, their formation took place as groups "anchored" *within* the G77 (Third World Network, 2012) – reflecting the continued default membership of the G77, and its flexibility of allowing subgroup membership.

Conclusion

Coalitions represent diverse positions on negotiation outcomes, but are not formed in an institutional vacuum, and instead are also shaped by the rules and practices of their institutional context. In discussing the broad pattern of how developing country negotiating coalitions have formed and emerged in the UNFCCC process, this chapter has argued that the historical legacies of the negotiating process have themselves shaped the manner and type of coalitions that have formed among developing countries: it matters that the climate negotiations were first organised via a UNGA mandate; it matters that the G77 was the first and default coalition "layer" for developing countries; it matters that the procedures of the negotiating process themselves favour organisation into formal coalitions. All this points to the importance of a close attention to the temporal sequence in which negotiations have unfolded as more than just a chronological retelling, but where this temporal sequencing also has causal significance for subsequent patterns of coalition formation and development. The choices of developing countries to form coalitions (or the lack of choice over some coalitions) and the resulting overlaps are as much shaped by these legacies of the institutional context as they are by real or perceived material interests on climate change at given moments in time.

This attention to timing and sequencing also contributes to a more nuanced understanding of the nature of North–South differences in international climate politics. Despite the material heterogeneity among developing countries, the South in climate politics has itself been constructed by the institution within which negotiations were organised, in the form of the G77. Even as newer coalitions form, they reproduce the institutional logics that treat the G77 as the umbrella group to which all belong. Overlaps become a way for the G77 to manage differences, through legitimising subgroup formation as formal coalitions. Nonetheless, this dynamic raises questions over whether and how such subgroups actually aid or complicate the negotiating process.

Finally, this chapter has also illustrated how historical institutionalist concepts such sequencing and layering, and political space can also provide insights into rationales for coalition formation, by regarding coalitions as institutions, rather than just alignments of interests. To understand the *evolution* of international climate politics from one moment to another is a relational task, and analysis through this temporal dimension can and should be further applied to understand the overall negotiating process or even the international climate architecture beyond the UNFCCC itself (e.g. Bernstein & van der Wen, 2017).

References

Ashe, J. W., van Lierop, R., & Cherian, A. (1999). The role of the Alliance of Small Island States (AOSIS) in the negotiation of the United Nations Framework Convention on Climate Change (UNFCCC). *Natural Resources Forum, 23*, 209–220.

Barnett, J. (2008). The worst of friends: OPEC and G-77 in the climate regime. *Global Environmental Politics, 8*(4), 1–8.

Bernstein, S., & van der Wen, H. (2017). Continuity and Change in Global Environmental Politics. In O. Fioretos (Ed.), *International Politics and Institutions in Time* (pp. 293–317). Oxford and New York: Oxford University Press.

Betzold, C. (2010). 'Borrowing' power to influence international negotiations: AOSIS in the climate change regime, 1990–1997. *Politics, 30*(3), 131–148.

Betzold, C., & Weiler, F. (2018). *Development Aid and Adaptation to Climate Change in Developing Countries*. Cham, Switzerland: Springer.

Blaxekjær, L. Ø. (2020). Diplomatic Learning and Trust: How the Cartagena Dialogue Brought UN Climate Negotiations Back on Track. In C. Klöck, P. Castro, F. Weiler, & L. Ø. Blaxekjær (Eds.), *Coalitions in the Climate Change Negotiations*. Abingdon: Routledge.

Blaxekjær, L. Ø., Lahn, B., Nielsen, T. D., Green-Weiskel, L., & Fang, F. (2020). The Narrative Position of the Like-Minded Developing Countries in Global Climate Negotiations. In C. Klöck, P. Castro, F. Weiler, & L. Ø. Blaxekjær (Eds.), *Coalitions in the Climate Negotiations*. Abingdon: Routledge.

Bodansky, D., Brunnée, J., & Rajamani, L. (2017). *International Climate Change Law*. Oxford: Oxford University Press.

Brazil. (2010, July 26). Joint statement issued at the conclusion of the fourth meeting of minsters of the BASIC group. Retrieved from http://www.itamaraty.gov.br/sala-de-imprensa/notas-a-imprensa/joint-statement-issued-at-the-conclusion-of-the-fourth-meeting-of-ministers-of-the-basic-group-rio-de-janeiro-25-26-july-2010.

Buys, P., Deichmann, U., Meisner, C., That, T. T., & Wheeler, D. (2009). Country stakes in climate negotiations: Two dimensions of vulnerability. *Climate Policy, 9*(3), 288–305.

Chan, N. (2013). *The construction of the South: Developing countries, coalition formation and the UN climate change negotiations, 1988–2012* (DPhil). University of Oxford, Oxford.

Chan, N. (2016a). Climate contributions and the Paris agreement: Fairness and equity in a bottom-up architecture. *Ethics & International Affairs, 30*(3), 291–301.

Chan, N. (2016b). The new impacts of the implementation of response measures. *Review of European, Comparative & International Environmental Law, 25*(2), 228–237.

Chasek, P. (2005). Margins of power: Coalition building and coalition maintenance of the South Pacific island states and the alliance of small island states. *Review of European Community and International Environmental Law, 14*(2), 125–137.

Dimitrov, R. (2016). The Paris agreement on climate change: Behind closed doors. *Global Environmental Politics, 16*(3), 1–11.

Eckersley, R. (2012). Moving forward in the climate negotiations: Multilateralism or minilateralism. *Global Environmental Politics, 12*(2), 24–42.

Edwards, G., Adarve, I. C., Bustos, M. C., & Roberts, J. T. (2017). Small group, big impact: How AILAC helped shape the Paris agreement. *Climate Policy, 17*(1), 71–85.

Farrell, H., & Finnemore, M. (2017). Global Institutions Without a Global State. In O. Fioretos (Ed.), *International Politics and Institutions in Time* (pp. 144–164). Oxford and New York: Oxford University Press.

Fialho, D. (2012). Altruism but not quite: The genesis of the least developed country category. *Third World Quarterly, 33*(5), 751–768.

Fioretos, O. (2011). Historical institutionalism in international relations. *International Organization, 65*(2), 367–399.

Fry, I. (2016). The Paris agreement: An insider's perspective—The role of small island developing states. *Environmental Policy and Law, 46*(2), 105–108.

Hallding, K., Jürisoo, M., Carson, M., & Atteridge, A. (2013). Rising powers: The evolving role of BASIC countries. *Climate Policy, 13*(5), 608–631.

Hanrieder, T., & Zürn, M. (2017). Reactive Sequences in Global Health Governance. In O. Fioretos (Ed.), *International Politics and Institutions in Time* (pp. 93–116). Oxford and New York: Oxford University Press.

Hochstetler, K. (2012). The G-77, BASIC and global climate governance: A new era in multilateral environmental negotiations. *Revista Brasileira de Política Internacional, 55*, 53–69.

Hoffmann, M. J. (2005). *Ozone Depletion and Climate Change*. Albany: SUNY Press.

Hopf, T. (2010). The logic of habit in international relations. *European Journal of International Relations, 16*, 539–561.

Mahoney, J., & Thelen, K. (2009). *Explaining Institutional Change*. Cambridge: Cambridge University Press.

Mathiesen, K., & Harvey, F. (2015, 8 December). Climate coalition breaks cover in Paris to push for binding and ambitious deal. *The Guardian*. Retrieved from https://www.theguardian.com/environment/2015/dec/08/coalition-paris-push-for-binding-ambitious-climate-change-deal.

Monheim, K. (2014). *How Effective Negotiation Management Promotes Multilateral Cooperation*. London: Routledge.

Mwandosya, M. (1996). *Survival Emissions: A Perspective from the South on Global Climate Change Negotiations*. Dar es Salaam: DUP and the Centre for Energy, Environment, Science and Technology.

Najam, A. (1994). The south in international environmental negotiations. *International Studies, 31*, 427–464.

Pierson, P. (2004). *Politics in Time*. Princeton: Princeton University Press.

Pistorius, T., Reinecke, S., & Carrapatso, A. (2017). A historical institutionalist view on merging LULUCF and REDD+ in a post-2020 climate agreement. *International Environmental Agreements: Politics, Law and Economics, 17*(5), 623–638.

Ramakrishna, K. (1992). Interest articulation and lawmaking in global warming negotiations: Perspectives from developing countries. *Transnational Law and Contemporary Problems, 2*, 153–172.

Roberts, J. T., & Parks, B. C. (2007). *A Climate of Injustice: Global Inequality, North-South Politics, and Climate Policy*. Cambridge, MA and London: MIT Press.

Roger, C., & Belliethathan, S. (2016). Africa in the global climate change negotiations. *International Environmental Agreements: Politics, Law and Economics, 16*(1), 91–108.

Sengupta, S. (2012). International Climate Negotiations and India's Role. In N. K. Dubash (Ed.), *Handbook of Climate Change and India* (pp. 101–117). Abingdon: Earthscan.

Sprinz, D., & Vaahtoranta, T. (1994). The interest-based explanation of international environmental policy. *International Organization, 48*(1), 77–105.

Third World Network. (2012). Meeting of the Like-Minded Developing Countries on Climate Change, Beijing, China, October 18–19, 2012. *TWN Info Service on Climate Change (Oct12/05)*. Penang, Malaysia: Third World Network.

United Nations. (1990). Protection of Global Climate for Present and Future Generations of Mankind. In *United Nations General Assembly Resolution A/RES/45/212*. New York: United Nations.

Vihma, A. (2015). Climate of consensus: Managing decision making in the UN climate change negotiations. *Review of European, Comparative & International Environmental Law, 24*(1), 58–68.

Vihma, A., Mulugetta, Y., & Karlsson-Vinkhuyzen, S. (2011). Negotiating solidarity? The G77 through the prism of climate change negotiations. *Global Change, Peace & Security, 23*(3), 315–334.

Watts, J. (2020). AILAC and ALBA: Differing Visions of Latin America in Climate Change Negotiations. In C. Klöck, P. Castro, F. Weiler, & L. Ø. Blaxekjær (Eds.), *Coalitions in the Climate Change Negotiations*. Abingdon: Routledge.

Weiss, T. G. (2009). Moving beyond north-south theatre. *Third World Quarterly, 30*(2), 271–284.

Williams, M. (1993). Re-articulating the third world coalition: The role of the environmental agenda. *Third World Quarterly, 14*(1), 7–29.

Williams, M. (1997). The group of 77 and global environmental politics. *Global Environmental Change, 7*(3), 295–298.

Williams, M. (2005). The third world and global environmental negotiations: Interests, institutions and ideas. *Global Environmental Politics, 5*(3), 48–69.

Yamin, F., & Depledge, J. (2004). *The International Climate Change Regime: A Guide to Rules, Institutions and Procedures*. Cambridge: Cambridge University Press.

Part II

Individual coalitions in the climate change negotiations

5   Pacific Island States and 30 Years of Global Climate Change Negotiations

George Carter

Introduction

Despite three decades of global negotiations under the United Nations Framework Convention on Climate Change (UNFCCC), there is still a divide on how to combat climate change. The time-consuming and arduous multilateral endeavour may seem to yield little reward; however, without this collective platform, there would be very few national and individual behavioural changes taken to avert or slow down the impacts of climate change. For none is this line of argument of "something is better than nothing" more relevant than for states from the Pacific, owing to their commitment to the UNFCCC process as a platform for advocacy. The 14 independent island states from the Pacific[1] have not only been seen and heard – they have been pivotal players in shaping the agenda and course of the climate change regime. Global climate change negotiations are one of the greatest legacies of island states from the Pacific to international politics.

In recent years a small but growing literature has focused its attention on better understanding the contributions of the states from the Pacific (Carter, 2015; de Agueda Corneloup & Mol, 2014; Denton, 2017; Ourbak & Magnan, 2018; Titifanue, Kant, Finau, & Tarai, 2017). For years scholars and media discourses have linked "climate change, or at least its impacts … with the small and apparently vulnerable nations of the Pacific" (Campbell & Barnett, 2010, p. 87). Low-lying coral atoll nations like Kiribati, the Marshall Islands, and Tuvalu, and in effect the coalitions they belong to, have been affectionately known as the "conscience" or the "moral conscience" of the negotiation (Davis, 1996; de Agueda Corneloup & Mol, 2014). Countries like Vanuatu, Tuvalu, or the Marshall Islands have been vocal and instrumental in the history of the climate change negotiations, from the formation of the Alliance of Small Island States (AOSIS) to the creation of the game-changing High Ambition Coalition (HAC) at the 2015 Paris Summit. Nevertheless, despite the growing attention to understanding how Pacific islands behave in climate change negotiations, we have yet to understand their contributions to climate change coalitions. To date, most research has examined islands in general, and their core coalition, AOSIS (Ashe, Lierop, & Cherian, 1999; Betzold, Castro, & Weiler, 2012; Chasek, 2005). Much less is known about how individual islands or island groups, like the Pacific islands, have shaped AOSIS and the climate negotiations.

The aim of this chapter is to piece together some of these studies and to re-trace the history of Pacific island states in the climate change negotiations. More specifically this chapter asks: what are the contributions of Pacific states in UNFCCC negotiations through their involvement in political groupings or inter-state coalitions? To explore the contributions of each individual Pacific state would be beyond the scope of this chapter. However, the aim here is to trace and explain cases and events where Pacific states utilised coalitions, and where they were chairs of coalitions. Drawing from a methodology using process tracing of historical narratives, the chapter complements textual analysis with global talanoa. Global talanoa (Carter, 2018) incorporates various methods from talanoa (cultural and empathetic interviews) and political ethnography, including personal observation and participation in Pacific delegations at various global, international, and regional climate meetings.

The contributions of Pacific states will be traced based through three key eras of the climate regime. First, the "early years" or the establishment of the climate regime; the chapter follows particular events from Malé to Geneva to the signing of the Convention in Rio (1988–1994). Central to the activities of Pacific states was the creation of a trans-regional coalition – AOSIS – an alliance that became the core coalition for all Pacific states. The chapter will then trace events in the "implementation phase" of the Convention and the Kyoto Protocol, from Berlin to Copenhagen to Warsaw (1997–2013). During this period, Pacific states joined, and in some instances led, numerous coalitions that arose around specific issues, or after the breakdown of diplomatic talks in Copenhagen and attempts to recover the regime. The final section traces activities leading to the Paris Agreement and beyond (2014–today). It is this period where particular states like the Marshall Islands, Tuvalu, and Fiji showcase their diplomatic finesse that underpins the emergence of a new regional coalition specific to the Pacific: Pacific Small Island Developing States (Pacific SIDS).

From Malé and Geneva to Rio: founding of the core coalition, AOSIS

The emergence of climate change in the global consciousness would not have happened without the foresight and the contributions of small island states from the Pacific Ocean. Although scientists in the 1970s forewarned of global warming from greenhouse gas emissions (Jäger & O'Riordan, 1996), it would take the dramatic series of flooding from storm surges, inundation, and sea-level rises in Maldives in the late 1980s to provoke action. These environmental hazards were not isolated occurrences in the Indian Ocean, but shared across many low-lying coral atoll island communities, such as those in Kiribati and Tuvalu in the Pacific. Together with leaders from Fiji, Tonga, and Vanuatu, they joined the government of Maldives in the first ever Small States Conference on Sea Level Rise in Malé in November 1989 (Government of Maldives, 1989). This monumental meeting brought together small island states from around the world for the first time to initiate a global campaign to combat climate change and put it on the international agenda.

The Malé Declaration on Global Warming and Sea Level Rise was the strategic document that came out of the 1989 conference. The declaration put forth a goal for its members: the need to create an international framework convention on climate change. This ambitious goal would become the legacy of small states, in a time when Cold War bipolar politics was dismantling. The wisdom of the island leaders was to foresee that a multipolar world and faith in multilateralism would better serve their interests amidst the devastating impacts of climate change. The issue needed to be supplemented as a key international agenda in the new arena of international politics, and to do so they needed to be vocal and be seen as setting the agenda.

Campaign from the outside: building the narrative

In order to be vocal and be seen, the Malé Declaration instigated "a campaign to increase awareness of the international community of the particular vulnerability of the small States to sea level rise" (Government of Maldives, 1989). The campaign heeded a call to be active from the outside, and a campaign for action from within. Since 1989, signatories of the Malé Declaration called upon the support of multilateral political networks starting from regional and international forums. At the regional level Maldives would garner the support of forums like the South Asian Association for Regional Cooperation (Government of Maldives, 1989). Pacific states like Kiribati, Tuvalu, and Fiji would elevate and bring prominence to their membership of the then-South Pacific Forum (now the Pacific Islands Forum) in prioritising climate change (PIF, 1988, 1989). As regional support grew, the demand for an international framework found favour within the Commonwealth's Heads of Government meeting in Malaysia 1989. In its Langkawi Declaration on the Environment, the 54 member countries of the Commonwealth called "for the early conclusion of an international convention to protect and conserve the global climate and, in this context, applaud[ed] the efforts of member governments to advance the negotiation of a framework convention under United Nations (UN) auspices" (Commonwealth Heads of Government Meeting, 1989).

While small states were building support on climate change, a complementary process started in late 1988 under various branches of the UN. The United National Environment Programme established the Intergovernmental Panel on Climate Change (IPCC), a body of scientists to study and bring forth the science to explain the occurrences of sea level rises, storm surges, droughts, and heatwaves. When the first IPCC report was tabled at the World Climate Conference (WCC) in Geneva in 1990, a renewed sense of political agency and urgency from the Pacific was both visible and loud in their call for international climate change action. In what one scholar calls Tuvalu's emergence as climate change *cause célèbre*, its then-Prime Minister Bikenibeu Paeniu famously stated the following:

> I can assure each and every one of you that I speak today from real experience because I live on one of the … smallest island groups in the Pacific. We are

therefore, along with others, extremely vulnerable to environmental hazards and the dangers of Greenhouse Effect and sea level rise. These are the problems which we have done the least to create but now threaten the very heart of our existence.

(Campbell & Barnett, 2010, p. 87)

The statement was made before a backdrop of both serendipitous leadership and strategic timing by leaders from the Pacific in a meeting from 29 October–7 November 1990.

A key mandate of the Malé Declaration was for its leaders and officials to work together and create the Malé Action Group. In using this political legitimacy, Tuvalu's Paeniu and Vanuatu's ambassador to the United Nations, Robert Van Lierop, along with officials from Indian, Caribbean, Mediterranean, and other Pacific Ocean states came together to share resources and speak as one. At the time, these 24 states from the Malé meeting realised their potential at the WCC; their shared collective positions from a coalition of island nations could have a loud and powerful voice. The trans-regional coalition would formally introduce itself as the Alliance of Small Island States (AOSIS),[2] name which is believed to have been coined by the Environment Minister Lincoln Myers of Trinidad and Tobago (Ronneberg, 2016).

Moreover, Paeniu's remarks gave a face to climate change, and showed Tuvalu and other Pacific islands to be the people most vulnerable to its adverse effects. Although this narrative negatively views them as victims, the narrative "vulnerable frontline states" was used strategically to safeguard the interest of island states in the climate regime. AOSIS membership recognised that beyond ensuring climate change as an international agenda, it needed to build a narrative that AOSIS was at the frontline and most vulnerable to climate impacts.

Campaign from within: AOSIS

By December 1990, the moment had arrived after international lobbying by island states and media networks: the UN General Assembly passed Resolution 45/212, sponsored by the island state of Malta. It called for an arrangement to establish the Intergovernmental Negotiating Committee (INC) to negotiate a framework and relevant instruments on climate change (UN General Assembly Resolution 45/212, 1990). From February 1991 through May 1992, five sessions of the INC were open to all member states of the UN to put forward positions on the text of the new global agreement. AOSIS did not disappear into the abyss as an ad-hoc coalition from the WCC in Geneva. In the midst of all states and traditional inter-state coalitions (in the UN system at the time), AOSIS was the new coalition at the negotiating table.

From the very first INC, the coalition of island states formalised its intent in the climate talks by insisting that the framework recognise and give special recognition to small island developing states (SIDS) and by attaining necessary representation for all its member countries (Ronneberg, 2016). Building on the success of the WCC the group elected Vanuatu's Ambassador Van Lierop as the inaugural chair, who passionately argued the case and used the term "front-line states" in climate change (Ronneberg, 2016). Throughout the preparatory phases and climate change negotiations for the 1992 Rio de Janeiro Earth Summit, Pacific countries within AOSIS successfully pushed for recognition of SIDS and their special circumstances within the Summit's Agenda 21. This "SIDS special consideration" would prove crucial as a safety valve for all future environmental and sustainable development conventions to consider the needs and the limited capacities of island nations (Chasek, 2005). However, in order to ensure that the agenda remained alive, Pacific states needed to be consistent and invest in a vehicle or core coalition group. AOSIS membership was mobilised and present in every room of both climate change and sustainable development negotiations.

By May 1992, the UN multilateral process succeeded in finding an agreement on the final text of the climate change convention. The convention would be one of three environment frameworks adopted at the Rio Earth summit in June of that year. Under the leadership of Van Lierop, AOSIS successfully negotiated that an additional seat on top of the predetermined ten regional seats of the Bureau (two from each of the five UN regions: Asia Pacific, Latin America, Africa, Western Europe, and Eastern Europe) be from SIDS, as the most vulnerable frontline states. Van Lierop was asked to join the INC Bureau as the small islands representative and was tasked with leading the negotiations in drafting the text and initial body of the Convention (Shibuya, 1996). Heralded as one of the first occasions that a Pacific state was given a leadership role in UN negotiations (Bodansky, 1994), to this day, the seat has always been occupied by a Pacific state. It would become one of the unwritten rules of AOSIS that someone from the Pacific would hold the SIDS seat, as the Caribbean states had secured a seat in the Latin America region.

AOSIS: the core coalition

From the various INC New York meetings to the first UNFCCC Conference of the Parties (COP) in Berlin in 1995, the coalition grew in capacity, finesse, and strategy. One of the most notable legacies of AOSIS was introducing the first ever draft protocol to the Climate Convention. This draft protocol stated that Annex I Parties to the protocol were to reduce their CO_2 emissions by 2005 to a level of at least 20% below that of 1990 (ENB, 1995); although that goal would later be watered down to 5%, AOSIS set the bar high by affirming an extreme to start negotiations. The coalition was influential in the early years of the regime by what Betzold calls "borrowing" power. She notes that AOSIS, and in advent Pacific states were able to influence negotiations by working with networks of third parties (NGOs and media), using context (normative reasoning), target (position and outcomes), and process (structure and timing) strategies (Betzold, 2010).

Ultimately, in the early years of the regime, Pacific states would come to find a home in the negotiations in AOSIS. Beyond the coalition meetings during the

INC or COP meetings, most importantly, the coalition would meet throughout the year in the various corridors of the UN headquarters in New York. Although AOSIS existed without any formal charter, or a regular budget, the coalition's work was primarily coordinated through its members' UN diplomatic missions in New York, which are actively involved in planning for, or in actual negotiations under the UNFCCC. Major policy decisions were taken at the ambassadorial-level plenary sessions (AOSIS, 2015a). The network of AOSIS to other like-minded coalitions of the global south provided a platform through which island negotiators could lobby for their concerns. This informal network of New York-based negotiators was built by frequent meetings and everyday exchanges in the UN halls and corridors. Moreover, these interactions would be formalised into shared positions through common membership of AOSIS members in the other global south coalitions, the G77 and China, and the Least Developing Countries (LDCs), as well as issue-specific coalitions like the Climate Vulnerable Forum (CVF) and the Coalition for Rainforest Nations (CfRN). As proposed by Castro and Klöck (2020), thus, overlapping memberships in multiple coalitions have offered opportunities for Pacific states to garner broader support for their interests and positions.

The AOSIS Chair has traditionally been a UN ambassador that would rotate every three years between the coalition's three key sub-regions: Pacific, Caribbean, and Africa-Indian Ocean-Mediterranean-South China Sea (AIMS). These three sub-regions would nominate one representative each to form the AOSIS Bureau or executive. In more recent times, the three representatives would be selected from either a regional political organisation representation roster, or by general agreement amongst UN ambassadors.

Despite the fact that to this day there is no formal charter or document that sets out its operations, the coalition is active and continues its work year in and year out based on the decision-making process of consensus. No decisions or positions undertaken by the coalition are taken by vote – there must be some form of consent or agreement by all. In effect, this internal politics of parties sourcing the consent from within their coalitions is a reflection of the consensus decision-making process in the climate regime.

From Berlin to Copenhagen and Warsaw: balancing multiple coalitions

AOSIS was not the only coalition that states from the Pacific have joined. However, AOSIS has remained the central coalescing home for all Pacific negotiators in the subsidiary body and COP meetings throughout the 30 years of climate negotiations. In the implementation years of the Convention and Kyoto Protocol from Berlin in 1995 to Copenhagen in 2009 and Warsaw in 2013, the coalition of small island states did not only condition the strategies, but saw the maturity of certain Pacific states and leaders in their consistent involvement in global climate change politics.

Pacific states' leadership in AOSIS

The most influential position in the coalition is the Chair. The preoccupation of UNFCCC procedures in giving the chair of key coalitions speaking time at the plenary, and most importantly, making them privy to many last-minute negotiations (always as part of the President's Friends of the Chair, see Chan, 2020) meant whoever was the chair of AOSIS held a position of leverage and influence. The Chair of AOSIS, or the country delegation of the Chair was also the secretariat. In order to prepare for UNFCCC meetings, the AOSIS secretariat organised a team of "coordinators" to speak as lead negotiators on the key positions of the group. The lead negotiators were pre-selected from the three sub-regional groupings by members for geographical balance – but more importantly for their technical knowledge on the issue. It was in these positions of coalition coordinators where NGO negotiators played a key role (as discussed below), while the Chair firmly remained a political position, with either the Ambassador, or if present, the Minister of the state chairing AOSIS.

There were five occasions when Pacific leaders and their support delegations led AOSIS and spearheaded specific issues in the climate regime. Some of these negotiators have since become influential political leaders that have changed the course of the climate regime. When Ambassador Van Lierop was chair of AOSIS, as noted earlier, he enshrined Vanuatu's legacy setting the agenda of the regime (Shibuya, 1996). During the leadership of Tuiloma Neroni Slade of Samoa (1997–2002) the coalition pursued hardline measures against Annex I polluters by strengthening and keeping alive the Kyoto Protocol (AOSIS 2015a). Just as the implementation phase of the Kyoto Protocol started in 2004, a new personality from the Pacific, Ambassador Enele Sopoaga of Tuvalu came to prominence (2005–2006). He would later become the country's prime minister and be instrumental in the Paris Agreement. He was staunch in his leadership calling out the alarming rise in global emissions and failure of mitigation efforts by Annex I countries. The issue on parity of mitigation and adaptation and the need for more support for developing countries, especially the least developed countries was central in Sopoaga's chairmanship (AOSIS 2015a).

In more recent times when Nauru chaired AOSIS from 2012–2014, Ambassador Marlene Moses' focus was on ensuring a long-term agreement to be reached in 2015 by emphasising the 1.5° long-term temperature goal and the issue of loss and damage[3] (AOSIS, 2015b). She led an ambitious post-COP15 (Copenhagen) campaign of limiting the global temperature increase to 1.5°C, while supporting up-scaling and access to climate finance, capacity building, and technology, as well as avenues for loss and damage as key issues for island states (AOSIS, 2015b). The issue of loss and damage has been a pillar argument of AOSIS ever since it was unsuccessfully introduced by Vanuatu's Van Lierop in the 1990s (Shibuya, 1996). The issue was vindicated two decades later with the establishment of the Warsaw International Mechanism for Loss and Damage associated with Climate Change Impacts (Loss and Damage Mechanism) established at the Warsaw COP19 in 2013. This mechanism would become Nauru's and one of AOSIS' key successes in the negotiations.

Pacific states and NGO networks in AOSIS

The makeup of state delegations to UNFCCC negotiations is multi-actor. It is not uncommon for states, especially from the Pacific, to include NGO representatives supporting or speaking on behalf of states in negotiations. In fact, from the early days of establishment to the very composition of AOSIS country negotiators, there has been a smorgasbord of NGO representatives within AOSIS. In the early days, lawyers from the Foundation for International Environmental Law and Development (FIELD) were part of Pacific state delegations that provided legal expertise. According to one negotiator, "AOSIS were united with FIELD lawyers on the complicated issues of markets and carbon trade that needed an expert approach".[4]

Scientific and environmental NGOs like Greenpeace and the World Wildlife Fund, as well as regional inter-governmental organisations like the Secretariat of the Pacific Regional Environment Program (SPREP) to name but a few, found a home in the coalition. In recent years, personnel from the Germany-based NGO Climate Analytics were embedded in various delegations from the Pacific, Caribbean, and AIMS countries, as well as in delegations from Africa, Latin America, Asia, and even the European Union. The NGO network of science strategists provided the essential research to back up the politics and moral position of AOSIS. The organisation would enact this in what they call "real-time support" science briefings during the negotiations, as well as negotiator capacity and ministerial support through regional training programs in preparing for UNFCCC meetings (Climate Analytics, 2020). So, while informing and supporting individual states and the AOSIS coalition, the network of scientists was also present in other coalitions.

Beyond legal and scientific support, NGOs would also play a vital back office role for the AOSIS secretariat or the Chair. During the Nauru AOSIS chairmanship of 2011–2014, Ambassador Moses affectionately called her team the "Pacific Chair" that comprised not only top women lead negotiators that were selected from around the Pacific, but also included the New York-based NGO Islands First. Islands First played a pivotal role in facilitating meetings and coordinating the AOSIS coordinators, lead negotiators, and the Chair. By providing these coalition secretariat services for Nauru and consecutive chairs of AOSIS like Maldives during the Paris Agreement negotiations, the NGO took the lead in legal support and media/public diplomacy initiatives. Non-state actors became essential in allowing leaders, diplomats, and officials from island nations to quickly learn "climate speak" but also to work strategically as a collective. As the next section discusses, while AOSIS is traditionally the main coalition body to which Pacific states are aligned, it is not the only political grouping in which they are active in the negotiations.

Pacific states in multiple coalitions

For small state delegations, association with and active participation in UNFCCC coalitions allow them to manage regime complexity (Chasek, 2005; Dupont,

1996), and in return act as coalescence vehicles to test and collect support for a louder voice. Among the records and literature of climate negotiations leading up to 2014, Pacific states since the beginning of the convention belonged to two coalitions, AOSIS and G77. As negotiations progressed over the years, some became active in the LDC group with similar country status, as well as initiating or joining other climate-specific coalitions: the CfRN and the CVF. The involvement of states from the Pacific in these coalitions is discussed in detail in the following paragraphs.

Group of 77 and China (G77)

Considered the hub of global south coalitions, the G77 has been in existence in most UN meetings and conventions since the establishment of the UN Conference on Trade and Development in 1964 (see also Chan, 2020). With 134 members, this dynamic group is the main clearing house for its members and their respective coalitions from various regions: African Group of Negotiators, Arab League, Latin American groups (Group of Latin America and the Caribbean – GRULAC, Bolivarian Alliance for the Peoples of Our America – ALBA, Independent Association of Latin America and Caribbean States – AILAC), small island states (AOSIS), LDCs, oil-exporting countries (Organization of Petroleum Exporting Countries – OPEC), and for large (emerging Brazil-South Africa-India-China – BASIC) and middle-income nations. The G77, like AOSIS, has been active since the early days of the UNFCCC. Samoa and Papua New Guinea under various occasions have been selected to coordinate auxiliary bodies of the coalition, but the first Pacific state to ever chair and convene the Group's positions in climate negotiations was Fiji in 2013. Fiji's chair in 2013 would prove timely in UNFCCC negotiations, as it would link to the year Nauru would lead AOSIS in securing the loss and damage issue under the Warsaw Mechanism, as discussed below. Despite being the largest grouping, only ten states[5] from the Pacific are part of the coalition. Cook Islands and Niue are not state members of the UN, while Palau withdrew from the coalition in 2004 when its president raised a growing gap between the needs of G77 and AOSIS (Campbell and Barnett (2010), 101) and no significant incentives and opportunities for Palau as a member (Pacific Islands Report, 2004). According to Campbell and Barnett (2010), Tuvalu has preferred not to join the group, citing differences in positions with bigger states in G77.

Least Developed Countries (LDCs)

The term "Least Developed Country" is a UN categorisation of countries from the 1970s, but the coalition itself became active in the climate negotiations in 2000. Concerned with the vulnerabilities and capacities of poorer countries to effectively report on requirements of the Kyoto Protocol, a coalition was created to ensure special consideration of LDCs from around the world (Chan, 2013). In a repeat of history as the initial head of AOSIS, Vanuatu would again become the inaugural chair of LDCs, when the coalition made submissions at subsidiary body meetings in 2000. Working together to defend the vulnerability of their states and claims to adaptation consideration (UNFCCC, 2017), LDCs have been instrumental in creating a special LDC Fund and in ensuring a permanent agenda in the subsidiary bodies to allow for discussion of special matters relating to LDCs. The LDC fund, on top of other UNFCCC mandated funds like the Green Climate Fund and the Global Environment Facility, provided the 48 member states, including the then-five members from the Pacific, with more direct climate finance access. Before Samoa graduated from LDC status in 2014, leaving Tuvalu, Solomon Islands, Kiribati, and Vanuatu in the coalition, it benefitted from high investment flows in its climate change projects especially through the LDC Fund. As one negotiator casually remarked, "Samoa had the brightest bunch, during LDC days they had the highest climate aid per capita".[6] In more recent years, Solomon Islands and Tuvalu have become key players in the coalition, especially securing funding for LDCs, and pursuing the issue of loss and damage. During the Paris COP, Tuvalu strategically utilised the coalition as its gateway to form a strong position with other members of the G77. Its collective position would force a bilateral negotiation with the USA in the final days of Paris.

Coalition for Rainforest Nations (CfRN)

As the regime moved into negotiations around carbon markets, an issue-specific coalition was born in 2004. The CfRN led by Pacific island state Papua New Guinea and other tropical rainforest developing countries emphasised the need to reconcile forest stewardship with economic development. Fiji, Samoa, Solomon Islands, and Vanuatu would become the other Pacific members of the coalition. The coalition operates as a forum to facilitate consensus among participating countries on issues related to domestic and international frameworks for rainforest management, biodiversity conservation, and climate stability. The work of the bloc has been instrumental in the establishment of the Reducing Emissions from Deforestation and Forest Degradation (REDD) program, which was vigorously negotiated as an insert to the 2007 Bali Action Plan. This success of the REDD program owes much to the tactical skill of Papua New Guinea's lead negotiator Kevin Conrad in the Bali COP of 2007. Famously quoted as the breakthrough moment in the "the mice that roared" incident, Conrad called out to the USA: "if you're not willing to lead, then get out of the way" (cited in Stiglitz, 2008).

Climate Vulnerable Forum (CVF)

This coalition, founded a month before COP15 Copenhagen in 2009, is an international partnership of countries highly vulnerable to a warming planet. The Forum claims to be based around the idea of a South–South cooperation platform for participating governments to act together to deal with global climate change (Climate Vulnerable Forum, 2016). Within its membership are the Marshall Islands, Kiribati, and Vanuatu. The forum of highly vulnerable developing countries meets within the margins of the negotiations in highlighting climate justice, human rights, and climate migration. Kiribati's former president Anote Tong was a key advocate of the coalition, and at one stage chaired various meetings of the group in Kiribati's capital Tarawa (Climate Vulnerable Forum, 2016).

Cartagena Dialogue for Progressive Action

After the breakdown of Copenhagen in 2009, a new group emerged in the following year committed to rebuilding trust between the developed and developing countries (Blaxekjær, 2020). The group continues to meet outside the formal UNFCCC negotiations. While its members claim it is not a political bloc, the dialogue provides a platform for delegates from developed and developing countries "to have frank discussions to better understand [each] others' positions and find areas of possible middle-ground" (Blaxekjær & Nielsen, 2015). Membership is fluid, as many developing countries try to avoid being associated too closely with developed countries due to a sense of loyalty to the G77 (Blaxekjær & Nielsen, 2015). Of the 30-plus countries involved in the dialogue, only Samoa and the Marshall Islands are from the Pacific. The 2014 Cartagena Dialogue was held in the Marshall Islands and focussed on opportunities to break the international deadlock and find common positions for a 2015 binding agreement (Islands Business, 2014).

Managing multiple coalitions

As more issues are being negotiated from the time of Berlin to Warsaw, there has also been a growth in the number of climate issue-specific coalitions (see Castro & Klöck, 2020). While it is the prerogative of states to choose to join a like-minded group that serves its interests, not all states have done so, and some have even opted to withdraw, as in the case of Palau and G77. Despite the growing complexity of the coalition landscape, AOSIS has remained the core coalition that facilitated the lobbying for the key demands of Pacific states. In part this was a result of the proud legacy of ownership – that Pacific states were a part of its establishment – with familiarity and relationships built over the years. Moreover, it is a procedural practicality. During a COP or subsidiary negotiations of the UNFCCC, a procedural allowance in the schedule would allow for a preparatory week for AOSIS states to meet first before meeting as G77. This preparatory week would allow negotiators to update and strategise in various coalition coordination groups. This practice would be replayed daily, where AOSIS negotiators meet as a group in the early hours of each morning, before meeting as G77 to refine and gather wider support before negotiations with Annex I states. The work and lobbying carried out by climate-specific coalitions like CVF or CfRN was to strengthen these positions in both AOSIS and G77 – again, membership of multiple coalitions therefore becomes an opportunity for Pacific states to gain more support for their positions (Castro & Klöck, 2020). However, if key demands from countries or climate-specific groups were not carried by the core coalitions, individual countries may choose to take the position directly, so as not to harm the collective integrity of the group. While this is an option, it has seldom been used, as most would choose to bring the position up again in future negotiations.

From Paris to beyond: evolution and Pacific finesse

There was heightened uncertainty and urgency in the years, and most notably the months, leading up to the 2015 Paris Agreement. With the world's attention fixed on whether or not there should be a new agreement to replace the Kyoto Protocol, one thing was certain: change was coming for the climate regime. In the end, the negotiations evolved into a bottom-up process based on voluntary contributions rather than a robust compliance system. During the Paris negotiations, the structures and strategies employed by these coalitions conditioned the way delegates behaved and carried out their work. Coherence amongst the coalition membership was highly valued; moreover, fragmentation into smaller coalitions was also highly probable. This fragmentation was evident in the emergence of a region-specific coalition, as the following describes.

Pacific SIDS coalition: capacity and networks

The Paris summit (COP21) would see the emergence of a Pacific SIDS coalition. For years, Pacific negotiators were already meeting informally on the fringes of the AOSIS and G77 meetings, sharing information on the various coalitions they were associated with. The coalitions, especially AOSIS, had served them well, yet the heightened urgency for stronger action in 2015 saw greater political involvement. Various regional political declarations, especially mandates from leaders at the Pacific Islands Development Forum and the Pacific Islands Forum called for Pacific negotiators and leaders to mobilise in the talks, and work as a unit. The mandate was an attempt to synchronise the work of diplomats based at their New York offices and their collaborative system of sharing and negotiating as a team, with technical experts working in the countries and regional organisations. The group did not develop from disagreements in the traditional coalitions, but rather from the shared realisation by leaders and negotiators from the Pacific that they would need to have a louder united voice in the negotiations.

With this, the Pacific SIDS membership agreed to establish four coordination groups, which would be the four key priorities for the coalition: a 1.5°C long-term temperature goal, adaptation, finance, and loss and damage. The new coalition continued to work within the auspices of AOSIS and other established coalitions. However, the fundamental task of Pacific SIDS was to act as the hub for collecting and sharing information in the numerous negotiation chambers, with the support of Pacific regional organisations, to empower both technical negotiators during the discussions, and the ministers in the Comité de Paris in the agreement phase of the negotiations.

An intrinsic feature of the Pacific SIDS coalition was its multi-actor team. The 14 Pacific delegations were small in numbers, with poorly resourced and under-trained officials. To overcome these capacity constraints, coalitions like Pacific

SIDS and AOSIS were heavily reliant on partnerships with networks that aligned them with regional organisations and NGOs. These non-state actors were able to successfully lobby for their objectives and influence the negotiations through timely research support. Most NGOs in the Pacific delegations were equally accorded official technical negotiators; and with the small size of delegations, most negotiators from the Pacific, whether official or NGO, were given a degree of freedom and autonomy to speak on behalf of states. Furthermore, since NGOs were not bound by state politics there was another advantage of using them in delegations, as one NGO negotiator said, "NGOs are *rat cunning*, they know how to cut corners and strategise with their networks in other delegations and outside in the media to get the work done".[7] Within Pacific SIDS and the events of COP25, two Pacific leaders and their state delegations would emerge to influence and shape the final outcome of the Paris Agreement: Tuvalu and the Marshall Islands.

Tuvalu's loss and damage, Marshall Islands' High Ambition Coalition

Although the issue of loss and damage was introduced unsuccessfully by Vanuatu in 1992, it was reinvigorated by Nauru under AOSIS in 2013 and with the support of Fiji during its leadership of G77, with the establishment of the Warsaw International Mechanism on Loss and Damage. When the issue was brought forth to be included in the new Paris Agreement negotiation, it would be Tuvalu's Prime Minister Enele Sopoaga who would become its champion. Tuvalu's delegation maintained its red line on the inclusion of loss and damage with a clause on compensation until the very last days of the talks (Fry, 2016). Its delegation throughout 2015 had successfully gathered enough momentum by working with its associated coalitions, the LDCs and the G77. As events unfolded, this red line would force the USA, who had different views on the issue, to reach out to Tuvalu for a last-minute bilateral compromise. Sopoaga's staunch strategy paid off. While allocations for compensation were removed, the issue of loss and damage would be anchored in Article 8 of the Paris Agreement.

Furthermore, Marshall Islands' Foreign Minister Tony deBrum was persistent in ensuring an ambitious agreement resting on the inclusion of 1.5°C in the long-term temperature goal. Unlike Tuvalu, deBrum reached out to ministers from the Global South, and across the firewall with the European Union and the USA. In secret informal ministerial meetings throughout the year, the Marshall Islands would build the trust and confidence of these leaders to form the High Ambition Coalition and strike a deal that would see them amenable to the inclusion of 1.5°C in the final outcome. Both ministers and their delegations were consistent in their participation and their messaging throughout the year. While one stood firm on their red line, the other forged partnerships; Tuvalu and Marshall Islands were small parties, but with big positions.

Fiji's Island COP: Talanoa Dialogue and oceans

Since its foundation in 2015, the Pacific SIDS coalition has become a robust negotiation bloc that has received both accolades and recognition within the UNFCCC negotiations. Regional organisations and their political declarations have identified the need to support the Pacific voice in climate change negotiations. Pacific SIDS were instrumental in supporting Fiji's bid to be the president of the UNFCCC COP23 in 2017. The year 2017–2018 saw evidence of the growing finesse of peoples from the most vulnerable region and their capacity to take charge in finding global solutions. Pacific states are no longer just able participants, but also able leaders in global climate change negotiations.

From Fiji's tenure at the highest position of the climate regime, stories of the vulnerabilities and resilience of island states were elevated in the COP23 Island COP. More notable were two key contributions, first including the protection of oceans in the working agenda of the climate negotiations. Parties and coalitions are now working towards solutions under the climate regime that incorporate protecting and enhancing the blue economy when considering actions around adaptation and mitigation. Second, through Fiji, the procedural work of the negotiations now incorporates the Talanoa Dialogue as a tool to facilitate discussions. Talanoa or the practice of inclusive, open, respectful, and non-confrontational dialogue was observed in the relationships within delegations, inside coalitions as well as inside negotiation chambers. This tool of diplomacy as practised in the regional political forums was elevated by Fiji to the global negotiations as part of the facilitative dialogue stocktake on contributions from parties since the Paris Agreement (Carter, 2017). The UNFCCC Talanoa Dialogue, which facilitates inclusive, open, and non-confrontational dialogue amongst state and non-state parties during the negotiations, has been embraced by negotiators and is expected to remain a permanent feature of global climate change negotiations.

Specific Pacific finesse

Since Paris, there has been resurgence in the way certain Pacific states and the Pacific region as a whole participate in the UNFCCC regime. While still maintaining allegiance to AOSIS and other coalitions, there has been an emphasis on more specific Pacific finesse or presence in the negotiations. The experience of Tuvalu shows it can manoeuvre negotiations with bigger states by utilising traditional coalitions like LDCs and G77 for numbers support. In the case of the Marshall Islands, they influenced negotiations by initiating a new coalition that extends beyond the North–South divide. For Fiji, their experience as head of G77 in 2013, as well as partnerships with AOSIS and Pacific SIDS coalitions to build a narrative as the first island state to chair a COP, would rewrite the history books of UNFCCC. More importantly, the formalisation of Pacific SIDS, a regional coalition that will act as the coalescing hub for Pacific members, would safeguard the shared interests of Pacific states in an increasingly complex negotiating landscape. As the regime evolves into a voluntary system, it is possible that states will also re-evaluate their association and the usefulness of multiple coalitions. It is expected that AOSIS will continue to serve the interests of small states, but more likely that regional coalitions (like the CARICOM for the Caribbean region) will have a role in putting pressure on other regions and state to do more.

Conclusion

What are the contributions of Pacific states to UNFCCC negotiations through their involvement in political groupings or inter-state coalitions? For almost three decades, Pacific states have engaged in the negotiations by establishing a core trans-regional coalition (AOSIS), balancing multiple issue-specific coalitions (G77, LDCs, CfRN, CVF, the Cartagena Dialogue, and HAC), and finally evolving to establishing a region-specific coalition (Pacific SIDS). The contributions of the Pacific states and AOSIS to both the climate regime and diplomacy as a whole is profound. As one scholar argues,

> if the small island states had not been actively engaged in UN climate negotiations, it could be argued that preventative action, including UN capacity building (scientific research and skill building) would not emphasize lowland and small island concerns and transformative procedures, and dispute resolution mechanism in the Convention and the Kyoto Protocol would not have been incorporated in these arguments.
>
> (Chasek, 2005)

In the early years of the Convention, we see the establishment of a core coalition, one that was trans-regional, with members from the Atlantic, Caribbean, Indian, Mediterranean, and Pacific oceans. Leaders from the Pacific were norm entrepreneurs in their attempts to build a campaign amongst the networks of partner states in the region through the Pacific Islands Forum and internationally with the Commonwealth of Nations. Visionary leaders like Paeniu of Tuvalu and Van Lierop of Vanuatu were instrumental in forging a coalition presence within the Convention, which became known as AOSIS. Beyond this, the coalition built a formidable narrative of vulnerable frontline states with the moral conscience that supported their case for special consideration.

In the implementation years from Berlin to Warsaw, Pacific states used AOSIS as the vehicle to influence key issues within the Kyoto Protocol, including mitigation efforts, adaptation, climate finance, and loss and damage, to name but a few. This was clearly evident on the four occasions when the Pacific nations of Vanuatu, Samoa, Tuvalu, and Nauru were chairs of AOSIS. Furthermore, during the implementation years, states from the Pacific reached out to established and climate issue-specific coalitions in the regime for further support for the positions pursued in AOSIS. Pacific states would come to manage membership in multiple coalitions: the G77, CfRN, LDCs, CVF, and the Cartagena Dialogue.

Moreover, in the years leading to Paris 2015 and thereafter, we see the evolution of the climate regime. The world witnessed the growing finesse of Pacific states, in the form of Tuvalu taking on the USA in Loss and Damage, the Marshall Islands in spearheading the High Ambition Coalition, and Fiji in presiding over COP23. States from the Pacific have come full circle in formalising a region-specific coalition. The Pacific SIDS continues to provide technical and strategic support for its members today.

In the global climate change negotiations, it was easy to assume that the smaller states like those of the Pacific would have little or no influence. Having limited resources and manpower in the negotiations, it was expected that like their "sinking islands", their delegations would drown in the sea of complex information and multiple fast-paced negotiations. However, as this chapter shows, after years of consistent hard work as well as strong manoeuvring, entrepreneurship, and trust in coalitions, Pacific states have and will continue to be pivotal players in the global climate change negotiations.

Notes

1 The 14 Pacific island states that are members of the UNFCCC are Cook Islands, Fiji, Kiribati, Marshall Islands, Federated States of Micronesia, Nauru, Niue, Palau, Papua New Guinea, Samoa, Solomon Islands, Tonga, Tuvalu, Vanuatu.
2 From the original 24 states that met in 1990, the AOSIS coalition has grown to include Antigua and Barbuda, Bahamas, Barbados, Belize, Cape Verde, Comoros, **Cook Islands**, Cuba, Dominica, Dominican Republic, **Fiji**, **Federated States of Micronesia**, Grenada, Guinea-Bissau, Guyana, Haiti, Jamaica, **Kiribati**, Maldives, **Marshall Islands**, Mauritius, **Nauru**, **Niue**, **Palau**, **Papua New Guinea**, **Samoa**, Singapore, Seychelles, Sao Tome and Principe, **Solomon Islands**, St. Kitts and Nevis, St. Lucia, St. Vincent and the Grenadines, Suriname, Timor-Leste, **Tonga**, Trinidad and Tobago, **Tuvalu**, **Vanuatu** (Pacific states are in bold).
3 The issue of Loss and Damage refers to the harms caused by climate change (whether sudden-onset events such as climate disasters or slow-onset processes such as sea level rise) on human and natural systems. Negotiations revolve around establishing liability and compensation. This is an issue pursued by AOSIS and LDCs, but resisted by developed countries.
4 Interview with AOSIS negotiator (Talanoa 13), 18 November 2015 (interviewer: G. Carter).
5 Pacific states belonging to the G77 are Fiji, Federated States of Micronesia, Kiribati, Marshall Islands, Nauru, Papua New Guinea, Samoa, Solomon Islands, Tonga, and Vanuatu.
6 Interview with Pacific negotiator (Talanoa 26), 7 June 2015 (interviewer: G. Carter).
7 Interview with Pacific negotiator (Talanoa 26), 7 June 2015 (interviewer: G. Carter).

References

AOSIS. (2015a). About AOSIS. Retrieved from http://aosis.org/about/.
AOSIS. (2015b). Alliance of Small Island States: *25 Years of Leadership at the United Nations*. New York: AOSIS.
Ashe, J. W., Lierop, R. V., & Cherian, A. (1999). The role of the Alliance of Small Island States (AOSIS) in the negotiation of the United Nations Framework Convention on Climate Change (UNFCCC). *Natural Resources Forum*, *23*(3), 209–220.
Betzold, C. (2010). 'Borrowing' power to influence international negotiations: AOSIS in the climate change regime, 1990–1997. *Politics*, *30*(3), 131–148.
Betzold, C., Castro, P., & Weiler, F. (2012). AOSIS in the UNFCCC negotiations: From unity to fragmentation? *Climate Policy*, *12*(5), 591–613.
Blaxekjær, L. Ø. (2020). Diplomatic Learning and Trust: How the Cartagena Dialogue Brought UN Climate Negotiations Back on Track. In C. Klöck, P. Castro, F. Weiler,

& L. Ø. Blaxekjær (Eds.), *Coalitions in the Climate Change Negotiations*. Abingdon: Routledge.

Blaxekjær, L. Ø., & Nielsen, T. D. (2015). Mapping the narrative positions of new political groups under the UNFCCC. *Climate Policy, 15*(6), 751–766.

Bodansky, D. (1994). *Prologue to the Climate Change Convention.* Paper presented at the Negotiating Climate Change: The Inside Story of the Rio Convention.

Campbell, J., & Barnett, J. (2010). *Climate Change and Small Island States: Power, Knowledge and the South Pacific.* London: Routledge.

Carter, G. (2015). Establishing a Pacific Voice in the Climate Change Negotiations. In G. Fry & S. Tarte (Eds.), *The New Pacific Diplomacy.* Canberra: ANU e-Press (pp. 205–218).

Carter, G. (2017). The island COP: Changing the negotiation climate with a 'Bula Spirit'. Retrieved from http://www.devpolicy.org/the-island-cop-changing-the-negotiation-climate-20171109/.

Carter, G. (2018). *Multilateral consensus decision making: How Pacific island states build and reach consensus in climate change negotiations* (Doctor of Philosophy). Australian National University Canberra.

Castro, P., & Klöck, C. (2020). Fragmentation in the Climate Change Negotiations: Taking Stock of the Evolving Coalition Dynamics. In C. Klöck, P. Castro, & F. Weiler (Eds.), *Coalitions in the Climate Change Negotiations*. Abingdon: Routledge.

Chan, N. (2013). *The construction of the South: Developing countries, coalition formation and the UN climate change negotiations, 1988–2012* (PhD thesis). Department of Politics and International Relations, University of Oxford, Oxford.

Chan, N. (2020). The Temporal Emergence of Developing Country Coalitions. In C. Klöck, P. Castro, F. Weiler, & L. Ø. Blaxekjær (Eds.), *Coalitions in the Climate Change Negotiations*. Abingdon: Routledge.

Chasek, P. S. (2005). Margins of power: Coalition building and coalition maintenance of the South Pacific island states and the alliance of small island states. *Review of European Community & International Environmental Law, 14*(2), 125–137.

Climate Vulnerable Forum. (2016). Climate vulnerable forum: About. Retrieved from https://thecvf.org/web/climate-vulnerable-forum/.

Climate Analytics. (2020) High-Level Support Mechanism for LDC and SIDS on Climate Change, retrieved from https://climateanalytics.org/projects/hlsm-high-level-support-mechanism-for-ldc-and-sids-on-climate-change/

Commonwealth Heads of Government Meeting. (1989). *Langkawi Declaration on the Environment*. Langdawi, Malaysia: Commonwealth.

Davis, W. J. (1996). The Alliance of Small Island States (AOSIS): The international conscience. Asia-Pacific Magazine, *2*(May), 17–22.

de Agueda Corneloup, I., & Mol, A. P. (2014). Small island developing states and international climate change negotiations: The power of moral "leadership". *International Environmental Agreements: Politics, Law and Economics, 14*(3), 281–297.

Denton, A. (2017). Voices for environmental action? Analyzing narrative in environmental governance networks in the Pacific Islands. *Global Environmental Change, 43*, 62–71.

Dupont, C. (1996). Negotiation as coalition building. *International Negotiation, 1*(1), 47–64.

ENB. (1995). *Summary of the First Conference of the Parties Framework Convention on Climate Change 28 March–7 April 1995*. Retrieved Berlin, from http://enb.iisd.org/vol12/1221000e.html.

Fry, I. (2016). The Paris agreement: An insider's perspective—The role of small island developing states. *Environmental Policy and Law*, *46*(2), 105.

Government of Maldives. (1989, 14–18 November). Conference Report. Paper presented at the Small States Conference on Sea Level Rise, Male.

Islands Business. (2014). Cartagena group expresses climate action optimism: Marshall Islands foreign minister. Retrieved from http://www.islandsbusiness.com/news/mars hall-islands/5038/cartagena-group-expresses-climate-action-optimism-/.

Jäger, J., & O'Riordan, T. (1996). The history of climate change science and politics. *Politics of Climate Change: A European Perspective*, 1–31. London: Routledge.

Ourbak, T., & Magnan, A. K. (2018). The Paris agreement and climate change negotiations: Small islands, big players. *Regional Environmental Change*, *18*(8), 2201–2207.

Pacific Islands Report,. (2004) Palau withdraws from UN Group of 77. Retrieved from http://www.pireport.org/articles/2004/11/24/palau-withdraws-un%E2%80%99s-group -seventy-seven

PIF. (1988). Forum Communique: Nineteenth South Pacific Forum Nukualofa Tonga. Nukualofa: Pacific Islands Forum.

PIF. (1989). Forum Communique: Twentieth South Pacific Forum Tarawa Kiribati. Tarawa: Pacific Islands Forum (South Pacific Forum).

Ronneberg, E. (2016). Small Islands and the Big Issue: Climate Change and the Role of the Alliance of Small Island States. In C. P. Carlarne, K. R. Gray, & R. Tarasofsky (Eds.), *The Oxford Handbook of International Climate Change Law* (pp. 761–778). Oxford: Oxford University Press.

Shibuya, E. (1996). "Roaring mice against the tide": The South Pacific islands and agenda-building on global warming. *Pacific Affairs*, *69*(4), 541–555.

Stiglitz, J. E. (Producer). (2008, 2016). Heroes of the environment 2008. *Time*. Retrieved from http://content.time.com/time/specials/packages/article/0,28804,1841778_18417 79_1841795,00.html.

Titifanue, J., Kant, R., Finau, G., & Tarai, J. (2017). Climate change advocacy in the Pacific: The role of information and communication technologies. *Pacific Journalism Review*, *23*(1), 133–149.

UN General Assembly Resolution 45/212. (1990). *Protection of Global Climate for Present and Future Generations of Mankind*. New York: UN General Assembly.

UNFCCC. (2017). Party groupings. *Process and Meetings*. Retrieved from https://unfccc. int/process/parties-non-party-stakeholders/parties/party-groupings.

6 Diplomatic learning and trust

How the Cartagena Dialogue brought UN climate negotiations back on track and helped deliver the Paris Agreement

Lau Øfjord Blaxekjær

Introduction[1]

As much as COP21 in 2015 and the Paris Agreement is highlighted as a success, COP15 in 2009 left many participants and observers disillusioned about inclusive multilateralism and prospects for COP16 low (e.g. Dimitrov, 2010; Elliott, 2013; ENB, 2009; Monheim, 2015). "The current situation in the UN multilateral process is worse than before Copenhagen. Failure to adopt the lightest possible non-binding declaration underscores the bleak prospects of the consensus-based UN process for responding to climate change" (Dimitrov, 2010, p. 22). Leading up to COP16, "key players [said] a further breakdown in fresh discussions this fortnight could spell the end of the UN multilateral negotiating process" (Willis, 2010), leading to climate minilateralism in forums like G8, G20, or G2 (USA and China) (Casey-Lefkowitz, 2010). However, COP16 adopted the Cancún Agreements and re-established some trust and legitimacy in the UNFCCC; COP17 decided on a second commitment period for the Kyoto Protocol and delivered a text known as the Durban Platform, the process to negotiate a global agreement in 2015 at COP21. COP18 managed to finish the old negotiations from COP13 (the Bali Action Plan) and set a more detailed timetable for the Durban Platform (Christoff & Eckersley, 2013, p. 118),[2] as well as formally adopting a second commitment period under the Kyoto Protocol. Negotiations were back on track, only to see COP19 and COP20 not progressing as much as many hoped. Nonetheless, COP21 delivered the Paris Agreement, under which all countries commit to (nationally determined) contributions, and did what COP15 failed to do.

In an analysis based on a large number of interviews with UNFCCC participants, Monheim (2015, pp. 29–62) argues that diplomatic efforts of the Mexican COP presidency and informal dialogues such as the Cartagena Dialogue for Progressive Action ("Cartagena" in the following) made the difference in getting negotiations back on track. Given the multiplication of coalitions (Castro & Klöck, 2020; Klöck, Weiler, Castro, & Blaxekjær, 2020), this chapter analyses Cartagena's role, thus contributing to a better understanding of the continuities and changes in the negotiation landscape leading up to the Paris Agreement.

Based on my own observations from COP17 to COP21 and 23 interviews with Cartagena participants (and research on coalitions in general), I demonstrate how Cartagena played an important role, first in getting negotiations back on track and since in delivering the Paris Agreement.

The chapter proceeds as follows: I start by contextualising Cartagena's emergence. In the following section, I present the research methodology and the 23 elite interviews and participant observations at COP17-COP21 that underpin the analysis. I also explain the concept of *communities of practice* that emerged during the research process as useful when analysing Cartagena. I discuss my results along three dimensions: *community* – who participates and how it evolved; *domain of knowledge* – the issues the participants share and work to resolve; and *practices* – participants' specific practices, or what they do and how. In the final section, I conclude that Cartagena as a community of practice got negotiations back on track through practices of boundary spanning, diplomatic learning, and trust. A central finding is that a practice I frame as *diplomatic learning* serves as an important contribution of Cartagena to its participants and to the process leading to the Paris Agreement.

Enter the Cartagena Dialogue for Progressive Action

Analysing climate diplomacy from the 1970s to the establishment of the UNFCCC and its many meetings since, Elliott (2013, p. 842) identifies "patterns that have persisted over time: last-minute diplomatic breakthroughs, moments of high acclaim usually followed by a tendency to fall back on 'talks about talks about talks', and constant struggles over technical and procedural details". Elliott (2013, p. 848) further notes that COP15 is not the only COP failing or almost failing, giving examples of COP2, COP6, and COP8. It seems the UNFCCC is designed to fail or only progress very slowly, first, because it is set up around climate clubs with professional negotiators defending history and national interests; and second, the principle of common but differentiated responsibilities and respective capabilities (CBDR-RC) has led to a stand-off between North and South (Elliott, 2013, p. 850f).

> [The North-South divide] would not necessarily be a problem, if there were sufficient trust to enable developed and developing countries to communicate positively and constructively. Unfortunately, however, negotiations between the groups tend to be dominated by knee-jerk suspicion, defensiveness, and misunderstanding, which hinder the rational discussion of proposals.
>
> (Depledge & Yamin, 2009, p. 444)

In short, COP15's failure could be ascribed to CBDR and the North–South divide with its lack of knowledge and trust between parties.

In the last-minute diplomatic breakthrough of COP15, many negotiators also had the experience that the big emitters USA and BASIC did not include others in the final negotiation (Elliott, 2013). BASIC was certainly criticised for this in G77 following COP15 (see Blaxekjær, Lahn, Nielsen, Green-Weiskel, & Fang, 2020). However, COP16 adopted the Cancún Agreements and re-established some trust and legitimacy in the UNFCCC. Elliott (2013, p. 847) cites then-UNFCCC Executive Secretary, Christiana Figueres, proclaiming at the closing ceremony that "the beacon of hope has been ignited and faith in the multilateral climate change process ... has been restored".

Blaxekjær and Nielsen (2015) argue that the changes in the organisational and narrative landscape of the climate negotiations since 2015 are important to understanding the road from Copenhagen to Paris. As this book also demonstrates, we have seen a proliferation of coalitions (Castro & Klöck, 2020), and the sharp division between developed and developing countries, the so-called firewall, has diminished. The Paris Agreement now supplements the principle of CBDR-RC with "in the light of national circumstances", meaning that all countries now have responsibility, because national circumstances change over time (UNFCCC, 2015).

Mexican climate diplomacy and Cartagena have been identified as key reasons for getting negotiations back on track. Cartagena is credited for substantial contributions to the Cancún Agreements and Durban Platform texts (Araya, 2011; Casey-Lefkowitz, 2010; Herold, Cames, & Cook, 2011; Herold, Cames, Cook, & Emele, 2012; Herold, Cames, Siemons, Emele, & Cook, 2013; Lynas, 2011). This is also part of the self-understanding of the participants, that Cartagena was instrumental in the successful outcome of COP16 "identifying areas of convergence and advocating ambition across its diverse regional constituencies. It is gearing up to similarly influence outcomes at COP 17 in 2011" (Commonwealth of Australia, 2011, p. 90).

Cartagena is framed as a *dialogue*, but what exactly does this mean? Climate diplomacy through *dialogue* is not a new phenomenon in UNFCCC negotiations. Examples include the "Greenland Dialogue on Climate Change" from 2005 to 2009 launched by the Danish COP15 presidency (Meilstrup, 2010, p. 120); the "Petersberg Dialogue" co-hosted by Germany and Mexico in 2009; the "Geneva Dialogue on Climate Finance" co-hosted by Mexico and Switzerland in 2009 (Monheim, 2015, pp. 50–51); or the "Toward 2015 Dialogue" convened by the Center for Climate and Energy Solutions (C2ES, 2015). These dialogues are high-level meetings. The Cartagena Dialogue is different and is better described as a coalition or a community of negotiators (Blaxekjær & Nielsen, 2015). What we know from public sources is that Cartagena was established in March 2010. Cartagena materialised from existing informal networks of experienced negotiators from Europe, the Alliance of Small Island States (AOSIS), least developed countries (LDCs), and Latin America and the Caribbean who, after a common experience of failure at COP15 and the feeling of being left out by the USA and BASIC, set up a first meeting in Cartagena, Colombia (Blaxekjær, 2016, p. 149; Blaxekjær & Nielsen, 2015). From early on, Cartagena and its participants understood this community as "an informal space, open to countries working towards an ambitious, comprehensive and legally-binding regime in the UNFCCC, and committed domestically to becoming or remaining low-carbon economies" (Lynas, 2011). The international diplomatic advisory group, Independent Diplomat, writes about its work with the Marshall Islands:

> [It] also included active leadership in, and technical support for, the emergence of the increasingly influential Cartagena Dialogue for Progressive Action. This new and unique grouping of about 30 progressive developed and developing countries from all regions of the world has worked to circumvent the north/south divide that has for too long characterised international diplomacy on climate change, and was crucial to achieving the key compromises that led to the much-needed Cancún Agreements at COP-16 in Mexico. After successful meetings in 2010 in Colombia, the Maldives and Costa Rica, the Cartagena Dialogue met for the first time in Malawi in early March 2011 to take further steps towards forging a new international consensus on climate change.
>
> (Independent Diplomat, 2015)

The Earth Negotiation Bulletin mentions Cartagena for the first time on 29 November 2010 at COP17:

> The third Cartagena Dialogue, an informal space open to countries working towards an ambitious, comprehensive and legally-binding regime under the UNFCCC, took place from 31 October – 2 November 2010 in San José, Costa Rica. It was attended by 29 parties from the Alliance of Small Island States, Latin America, Europe, Oceania, South East Asia and Africa. Participants reaffirmed their desire for an integrated and ratifiable post-2012 legal regime. They identified the need for substantial progress at COP 16, in the form of balanced decisions, to provide a foundation for this overarching objective. Participants also exchanged views on textual proposals.
>
> (ENB, 2010, p. 2)

Although Cartagena is not an official negotiation group, it seems to meet several times a year, exchange views on negotiation text, and seek a middle ground and an ambitious climate agreement. It has grown from 30 to at least 42 countries as described by the UNFCCC:

> A collection of around 40 countries working towards an ambitious legally binding agreement under the UNFCCC, and who are committed to becoming or remaining low carbon domestically. Participates include: Antigua & Barbuda, Australia, Bangladesh, Barbados, Burundi, Chile, Colombia, Costa Rica, Denmark, Dominican Republic, Ethiopia, European Union, France, Gambia, Georgia, Germany, Ghana, Grenada, Guatemala, Indonesia, Kenya, Lebanon, Malawi, Maldives, Marshall Islands, México, Netherlands, New Zealand, Norway, Panama, Peru, Rwanda, Samoa, Spain, Swaziland,

Sweden, Switzerland, Tajikistan, Tanzania, Uganda, UAE, and the United Kingdom.

(UNFCCC, 2014b)[3]

Based on what we know so far, and according to the coalition criteria in Castro and Klöck (2020) – *geographic and thematic scope, membership,* and *level of formality* – Cartagena is classified as a global issue-specific coalition with common objectives and values working informally yet in a coordinated manner through meetings between and during UNFCCC COPs and intersessionals. To learn more about how exactly Cartagena is different from other dialogues and coalitions and how it made an impact on climate negotiations, I first explain how I approached the study of Cartagena, and then present the analysis and findings.

Methodology and analytical concepts

I first learned about Cartagena at a climate policy workshop in January 2012, where a policy advisor on a developing country delegation gave a presentation. The policy advisor was one of the people setting up and developing Cartagena, and argued that this kind of dialogue and exchange of views brought climate negotiations back on track. I got a list of countries, so I could begin to examine it myself. I followed up and interviewed this policy advisor two years later (Table 6.1, Interview 14). Getting direct access to Cartagena practices will probably not be possible for any study of Cartagena for those other than insiders, first because of the time and geographical span of meetings and communication, and second, because most meetings and other activities are not public. As with other studies of diplomacy, studying what happens at closed meetings and other behind-the-scenes practices need to be interpreted from and through other sources, such as interviewing people who were there. "The rationale is that, even when practices cannot be 'seen', they may be 'talked about' through interviews or 'read' thanks to textual analysis" (Pouliot, 2013, p. 49). Beginning the research with the very little information available in 2012, I followed an abductive research methodology or abductive reasoning (Van de Ven, 2007, p. 20, 140f) and decided to conduct interviews and do participant observations at the following COPs and in that process identify and combine relevant theoretical concept to analyse what Cartagena does and how – as a continuous (abductive) dialogue with sources and concepts. This section presents this methodology and concepts in a more schematic way than the actual process.

I conducted 23 interviews from November 2012 to December 2015 (see Table 6.1). Interviews span all regional and group representation in Cartagena: 12 from developing countries and 11 from developed countries, covering ministers, negotiators, and policy experts on delegations. Interviews were granted based on anonymity. They took place at COPs, at a climate-related event in Copenhagen, over email, or phone. I consider 23 interviews in combination with COP observations and other first-hand sources relating to Cartagena to be sufficient. Interviewees gave similar answers, indicating that additional interviews would not provide much additional information. Interviewees emphasised different aspects relating to their personal experiences, which indicate that answers were not rehearsed and officially sanctioned. The same semi-structured interview guide was used for all interviews, with a focus on getting interviewees to describe the Cartagena history and purpose, Cartagena's *modus operandi* as well as its strengths, successes, and challenges.

Getting access to negotiators for interviews is difficult. Negotiators are busy with meetings and preparations and have little time for interviews. Further, the negotiation schedule changes all the time, and many interviews were either rescheduled or cancelled. Two strategies worked best: waiting outside the delegation or meeting rooms before or after scheduled meetings, and spotting the relevant negotiators to try to fix an interview, then persistently following up, and getting interviewees or other contacts working with potential interviewees to introduce me. Other primary and first-hand sources are a few media sources and press releases in relation to Cartagena meetings, a video-blog from the Norwegian delegation from COP19 explaining what Cartagena is, and observations from the Danish delegation's closed daily briefings at COP17 to COP21 (briefings are only used as background knowledge).

I also rely on background knowledge of the organisational developments stemming from participant observations at COP17 to COP21, and the intersessional in Bonn, June 2014, where I focused on negotiations on the Kyoto Protocol (KP), Long-Term Cooperative Action (LCA), and the Durban Platform for Enhanced Action (ADP). Specifically for my research on Cartagena, I attended as many plenary sessions on KP, LCA, and ADP as possible, where I took note of which countries spoke when and how statements related to each other. These observations confirmed what many interviewees said: that the same words or phrases discussed in Cartagena would reappear in negotiation interventions by Cartagena countries.

A lack in my approach is direct observations from Cartagena meetings. I asked a few times for permission to sit in on COP coordination meetings, but requests were denied. Even if granted permission to participate in a Cartagena meeting between sessions, it would not have been financially and practically possible for me to do so. Despite this lack, I have gained much new knowledge about Cartagena.

To analyse Cartagena, I first considered the concept of discourse coalition (Hajer, 1995). However, the focus would be on text and statements, thus missing how Cartagena is organised and works; an analysis would include text and statements by all countries; and a discourse coalition would likely include other countries than those in Cartagena, and some in Cartagena might not be included. Cartagena could also be analysed as an epistemic community (Adler & Haas, 1992; Haas, 1992), but from my interviews it became clear that Cartagena is not just about sharing knowledge and developing policy ideas, Cartagena is as much about the social interaction, shared values, and a feeling of belonging. Adler and Pouliot (2011), among others, recently developed a new perspective on international relations called *international practice*, in which they had incorporated the concept of *communities of practice* based on the work by Wenger, McDermott, and Snyder (2002).

Communities of practice

The concept of communities of practice is popular in many academic fields and diverges in four main approaches (for a critical discussion, see Cox, 2005). The concept applied here is based on the adaptation by Adler and Pouliot (2011) following Wenger et al. (2002), except for the part providing a manual for managers to set up communities of practice. The concept offers a focus on actors' shared practices, not just knowledge (Sebenius, 1992), but the continued interactions and mutually recognised performances of delegates (Adler, 2008; Adler & Barnett, 1998; Adler & Pouliot, 2011; Adler-Nissen, 2014; Neumann, 2002; Pouliot, 2008, 2010). Communities of practice are broadly defined as "groups of people who share a concern, a set of problems, or a passion about a topic, and who deepen their knowledge and expertise in this area by interacting on an ongoing basis" (Wenger et al., 2002, p. 4).

> The community of practice concept encompasses not only the conscious and discursive dimensions, and the actual doing of social change, but also the social space where structure and agency overlap and where knowledge, power, and community intersect. Communities of practice are intersubjective social structures that constitute the normative and epistemic ground for action, but they also are agents, made up of real people, who – working via network channels, across national borders, across organisational divides, and in the halls of government – affect political, economic, and social events.
>
> (Adler & Pouliot, 2011, pp. 17–18)

Communities of practice vary a great deal; they can be small or large (membership size), long-lived or short-lived, co-located or distributed (place of interaction), homogenous or heterogeneous (members' backgrounds), inside and/or across organisational boundaries, unrecognised or formally institutionalised, and can emerge spontaneously or intentionally (Wenger et al., 2002, pp. 24–27). Common to communities of practice is that they consist of a basic structure defined as 1) a *community* of people who care about 2) a *domain* of knowledge, which defines a set of issues, and 3) the shared *practices* that they are developing to be effective in their domain. To identify a community of practice one has to go beyond the abstract and desk-based studies: "You have to look at how the group functions and how it combines all three elements of domain, community, and practice" (Wenger et al., 2002, p. 44).

The *community* element is more than a description of the participants, their backgrounds, the size, meeting places, and frequency of meeting of the community.

In combination with the domain of knowledge it is about learning together and "a sense of belonging and mutual commitment" e.g. bringing down barriers and making it easier and more acceptable to approach each other and ask (the difficult) questions, ask for help, or voice ideas. People stay and contribute to this learning because of a recognised reciprocity, trust, and, with time, a developed shared history and identity.[4] However, communities also encourage and even thrive on differentiation, so individuals can develop own and distinct roles and contributions, fostering "richer learning, more interesting relationships, and increased creativity" (Wenger et al., 2002, p. 33ff).

The element of *domain of knowledge* is more than the common interest in an issue. It is the "raison d'être" of the community and what "guides learning". Members *know* what is relevant and useful for the community; they *experience* the same issues and problems; they share a *passion* and a *will* to collectively succeed and steward the community's expertise and capabilities. The community's expertise and capabilities are important vis-à-vis the wider organisational context. If recognised for its knowledge and developed practices, higher levels of authority in the field will most likely consult the community before making important decisions that relate to the community's knowledge domain (Wenger et al., 2002, p. 29ff).

Practices are what bind the community together and what e.g. makes it different from an epistemic community, analytically and practically, because of the focus not just on shared knowledge, but on the specific ways of doing things: "a set of common approaches and shared standards that create the basis for action, communication, problem solving, performance, and accountability" (Wenger et al., 2002, p. 37ff). Practices are based on both explicit and tacit knowledge, "a specialised tool or a manual, to less tangible displays of competence" like the ability to interpret slight changes in e.g. diplomatic language. Practices embody "a certain way of behaving, a perspective on problems and ideas, a thinking style, and even in many cases an ethical stance" (Wenger et al., 2002, p. 37ff). To understand change, we should study practices, because they shape "thought and language into regular patterns of performance and turn contexts or structures into (individual and corporate) agents' dispositions and expectations" (Adler & Pouliot, 2011, p. 20).

Analysing the Cartagena Dialogue for Progressive Action

The analysis follows the concepts of *community*, *domain*, and *practice* with most weight on the specific practices of Cartagena.

The Cartagena community

Cartagena really fits the definition of a community of practice. Interviewees all told similar stories of how it is a community of learning together: a community that had slowly developed from an already existing informal community of practice to an increasingly organised community responding to the failure of COP15 (and COPs before that) to genuinely listen and learn across the formal climate clubs and North–South divide.

> The Cartagena Dialogue is the only forum which crosses the developed-developing divide in the UNFCCC negotiations and includes participants from such diverse regions. ... The Cartagena Dialogue allows Parties to better understand others' positions and what is important to them, to better identify where common ground can be reached. Its value is also in the relationships and networks it forms with delegates from a diverse range of countries that might not ordinarily work together. Although participants see the value of the Dialogue, it is also difficult to pinpoint exact outcomes attributable to the Cartagena Dialogue. Recently the Cartagena Dialogue has considered more diverse issues, which makes it more difficult to find areas of common ground. Nonetheless, the Cartagena Dialogue is united by the desire for a global climate change agreement that keeps warming below two degrees, so it has the opportunity to build momentum and be influential in the 2015 agreement negotiations.
>
> (Table 6.1, Interview 15)

As the quotation suggests, Cartagena is a community that breaks down the firewall that other developing countries seek to uphold, but it also bridges the other major groups (see e.g. Depledge & Yamin, 2009; Yamin & Depledge, 2004). Cartagena has grown to regularly include 42 Parties and spans the Umbrella Group, the European Union, AOSIS, LDCs, the African Group of Negotiators (AGN), the Environmental Integrity Group (EIG), the Independent Association of Latin American and the Caribbean (AILAC), and the Group of 77 and China (G77), and thus many different cultural and political backgrounds. The regular participants (individuals) in Cartagena are senior and junior negotiators; ministers participated in meetings the first year, but rarely after that, although they were at the table during COP21's second week. These negotiators have been part of the UNFCCC negotiations for some time, thereby sharing a certain international culture focused on their special knowledge of negotiation issues. Theoretically, this shared knowledge and culture, but different backgrounds, are important to get a community to work in practice, because learning requires openness (Wenger et al., 2002).

Cartagena bridges other groups, which many interviewees mention as a reason for keeping Cartagena informal and not making statements: "[in 2010] Cartagena made a statement in plenary, which was not perceived as something positive by the rest of the G77" (Table 6.1, Interview 11). Both developed and developing country negotiators shared the observation that Cartagena as a *dialogue* is more acceptable to G77 and other groups because it is not a formal coalition. It is more practical for Cartagena to be an informal space distanced from formal political positions to be better able to explore new suggestions without being held responsible afterwards for what was said and written in an attempt to reach compromises.

This informality brings negotiations forward, because as many interviewees said, the ideas and knowledge created through Cartagena shape individual Parties' positions and proposals. Although informal, Cartagena is very organised with a secretariat and logistics team, which was run by the UK together with Australia from 2010 to 2013 and New Zealand from 2014. Since the end of 2013, the Australian government has abandoned progressive climate action. Then-Prime Minister Tony Abbott stated that "climate change science is crap" (cited in Marks, 2014; see also Readfearn, 2014), and at a concurrent Cartagena meeting, host and Foreign Minister of the Marshall Islands, the late Tony deBrum, expressed concern over Australia's commitment: "The previous government of Australia [was] instrumental in helping establish Cartagena Dialogue. This week, they're sending a very junior official to represent Australia. I'm not sure how we should interpret that" (cited in ABC, 2014). From 2014 Australia was no longer active in Cartagena (Table 6.1, Interview 17).

Cartagena has a core group of countries representing regional views. They prepare meetings, usually three a year prior to UNFCCC meetings, and lead groups preparing ideas and discussion papers. Work and meetings are usually financed on a project basis by developed countries. The internal organisation reflects, to the extent possible, views from different regions, both from developed and developing countries. It seems Cartagena participants have agreed to keep the size at 42 for it to be manageable, although one interviewee mentions "more than 50 Parties and more wanting to join" (Table 6.1, Interview 10). Australia is still listed. Others said that including more Parties is thought to jeopardise the personal and intimate atmosphere from years of regular meetings and discussions. In short, size is related to the level of trust between participants.

Domain of knowledge

The domain of knowledge for Cartagena can be divided into a more general level of Cartagena participants' shared understanding and experiences of what is at stake in UNFCCC negotiations and a more specific level of joint knowledge production and issues that Cartagena seeks to advance. Interviewees agree that at a general level, the Cartagena community shares an understanding that COP15 could have succeeded if Parties had actually listened to each other, and that it is of utmost importance that negotiations do not fail again. The Cartagena community is (self-)framed as taking "progressive action", understood to be in relation to advancing negotiations, creating a middle-ground, and exploring new ideas.

> We believe in 'shared responsibility and different responsibility'. We are more flexible on the Convention's principles, because the world has changed and is changing. And all are vulnerable, so to us, this principle or the principle of special circumstances should not be used as a way to keep from acting
>
> (Table 6.1, Interview 3)

Interviewees also point out that their countries have ambitious climate policies:

> [Participants] bring their country's positions and perspectives of course, but also expertise, experience from their national implementation and strategic views on how to make progress in the negotiations. The approach is very much one where we try to understand the basis for others' positions, to discuss ideas and ways to make progress, to discuss strategies.
>
> (Table 6.1, Interview 13)

At a more specific level, Cartagena's domain of knowledge mirrors UNFCCC negotiation issues, and these are organised in working groups or teams for each issue with a lead from a developed and developing country. One interviewee explained that lengthy analyses are prepared for and discussed at the Cartagena meetings between sessions, and then used at the COP and intersessional meetings (Table 6.1, Interview 11). However, discussions do not always lead to common ground. Cartagena has not been able to resolve differences between Parties on loss and damage, and some developing country participants are at times frustrated with developed countries' positions on finance.

Practices

I have identified a range of practices that Cartagena participants have developed over the years. For analytical purposes I have structured these practices under the following types: *boundary spanning*[5] (bridging formal groups), *face-to-face diplomacy* (meeting and engaging in dialogue), *internal and external communication*, and organising and advancing knowledge (or *diplomatic learning*).

Boundary spanning refers to practices where Cartagena participants from different political and regional groups come together, spanning some of the dividing organisational boundaries in UNFCCC negotiations. Boundary spanning occurs across the North–South divide, but also between regions, and types of Parties with different domestic and international experiences and knowledge. For example, delegates from Norway, Australia, and New Zealand are able to apply their knowledge from Umbrella Group meetings and better explain nuances of the American or Canadian positions to some developing countries which otherwise would get these "analyses of positions" from e.g. larger countries in G77. Likewise, developing country participants can explain to developed country counterparts where and how G77, AOSIS, LDCs, and AILAC could be open to compromises in negotiations. Furthermore, participants can share national experiences with policy planning and implementation, and how it can be incorporated in specific UNFCCC issues e.g. Nationally Appropriate Mitigation Commitments or Actions by Developed Country Parties and Nationally Appropriate Mitigation Actions by Developing Countries, Technology Transfer, or Capacity Building. Participants also share strategic views on how to make progress in the negotiations. Because

Cartagena builds on a culture of trust and knowledge of each other's positions and realities, participants are better able to know what relevant information to bring into the dialogue.

Face-to-face diplomacy refers to the actual meeting and engaging in dialogue (together with boundary spanning). One interviewee remarked that it is "the small stuff that is much more important" (Table 6.1, Interview 5). This "small stuff" happens through a range of practices: different types of meetings during or between UNFCCC meetings, preparatory meetings with lead groups or Cartagena secretariat, social meetings, and ad-hoc issue-specific meetings and huddles during UNFCCC negotiations. It is a *learning by doing* process, and Cartagena increasingly institutionalises its face-to-face diplomatic practices. Since 2010, Cartagena has been meeting regularly before UNFCCC meetings two to three times a year, always in a developing country. The host country sometimes invites neighbours, and the UNFCCC secretariat often participates or gives a video presentation. "At our meetings we have statements from the UNFCCC secretariat. These statements show the importance that UNFCCC secretariat attach to this group" (Table 6.1, Interview 10). In the first years, the host country had many tasks besides preparing the agenda e.g. distributing input from participants between meetings and during the COP. In 2012, Cartagena decided to establish a core group of countries with regional and political group representatives to support the host country, and to secure better inclusion of all views and ownership of the processes. This core group contributes to logistical decisions and assists the host country in preparing the meeting agenda. The host country is also a member of the core group prior and after the meeting to facilitate learning.

Practices at meetings include discussing and planning common approaches to upcoming UNFCCC negotiations issues (also based on sharing interpretations of previous negotiation outcomes), and also building stronger personal ties through social events like dinners where participants get to know each other better. The semi-planned social events are important to participants and something they look forward to, and also something that provides participants with the necessary confidence to approach each other during UNFCCC negotiation meetings outside of the Cartagena setting if a specific question arises that needs discussing. Time is short and some delegates are hard to locate since they do not have official office space. Knowing who to talk to from other delegations – especially across boundaries – is conducive to even attempting to do so. Cartagena delivers an open space for diplomatic learning, especially helpful for negotiators from developing countries without the same capacities as developed countries. This kind of enduring dialogue in practice throughout the year, at political, technical, practical, and personal level, is a significant process innovation in the UNFCCC field compared to e.g. the one-off high-level dialogues.

Cartagena invites observers to participate in their meetings. It is a common practice in negotiations for the political groups to set up meetings to explore positions or get input from NGOs and experts. Conference diplomacy can be understood as a collection of meetings. And it is of course one of the ways Cartagena participants can gain influence in negotiations, by engaging with other actors, especially the UNFCCC secretariat, the COP presidency, or specific negotiation issue co-chairs, like the ADP co-chairs who are tasked with trying to find common ground. Cartagena is an attractive negotiation partner because Cartagena has put in many efforts to explore common ground across North–South and regions on dividing issues. Many interviewees highlighted that Cartagena's negotiation input has bridging potential internally at both negotiators' level and ministers' level. Bridging also goes beyond Cartagena and the UNFCCC secretariat. One interviewee mentioned that Cartagena carefully crafts and fine-tunes messages, which helped the Mexican Presidency and "for example the ADP outcome is very much driven by the way we approached issues in Cartagena" (Table 6.1, Interview 5). Looking towards Paris, another interviewee explained:

> one of our strategies for COP21 is to engage, to have high-level ministerial bridging and we are developing messages that we are going to give, not by ourselves, but feed into the COP process, so that this messaging will actually …, we will have it at the high-level sessions.
>
> (Table 6.1, Interview 10)

Cartagena meets regularly during the UNFCCC meetings, sometimes twice a day; however, it can be difficult for all to participate, especially developing countries with small delegations that have to prioritise more official meetings over informal dialogue. However, being part of Cartagena and having developed personal relationships and knowledge of each other, many delegates also meet on their own. This type of ad-hoc face-to-face meeting rarely took place before COP15, but it is now common among Cartagena participants – a significant process innovation.

> It is a way of knowing what the other members think on specific issues and negotiation developments. We are more frank with each other and we don't negotiate – we share our views – and then we try to take other members' views into consideration. Because we ultimately want the same thing. This is a new way of demonstrating that it is possible to talk and work together between the developed and developing countries. The old North-South divide is based on narrow national interests and also old-fashioned ideologies.
>
> (Table 6.1, Interview 3)

Internal and external communication refers to the way Cartagena communicates. Internally, Cartagena communicates in an open and bridge-building style, and participants generally think of Cartagena as a space of sincere communication. (Of course, participants also communicate tactically, but the point is really to communicate and explain national positions.) Internally, Cartagena bases communication on a few principles: the Chatham House rule, presentation of as many views as possible, strong personal relationships, and no journalists present or reporting from meetings. Participants recognise that this space of sincere communication is something unique in UNFCCC negotiations that are usually based on rhetorical positions of national interests. This open, honest, relaxed, friendly space that allows participants from different groups to venture beyond national interests, to explore new ideas, and to figure out a way of creating common ground is central when interviewees talk about how Cartagena helped UNFCCC negotiations get back on track. Several interviewees mention that Cartagena ideas actually feed into Party submissions, statements, and the final UNFCCC agreements, especially in Cancún, Durban, and Paris.

Cartagena as a *dialogue* is different from UNFCCC *negotiations*, because Cartagena meetings do not have to result in a consensus agreement. Participants share and discuss views and can then incorporate these views or new common perspectives in their own national position or in formal groups as they find appropriate. Participants mention that they do learn something new, that meetings create added value. For specific issues, appointed leads will have prepared a presentation for discussion, but otherwise each participant is expected to share views and analyses for all to discuss. Outside of meetings, communication is based on an email listserv run by the secretariat as well as direct, personal communication between two or more participants e.g. in working groups over phone conference.

Although Cartagena engages and invites others, Cartagena does not engage in formal communication externally, like the Like-Minded Developing Countries (LMDC), BASIC, or AILAC. Cartagena participants have only made a few statements referring to Cartagena. Cartagena also rarely communicates through the media. There are some examples of media pieces, mostly in conjunction with the preparatory meetings in developing countries, where the host (often at ministerial level) speaks in general terms about the need for climate action, about the vulnerability of the host country and the region, and that progressive countries will meet there to prepare for the next UNFCCC meeting (IISD, 2012, 2014; Maldives, 2010a, b; RTCC, 2014). Some participants use their ministries' official channels to send a message of active engagement and bridge-building in UNFCCC negotiations (Commonwealth of Australia, 2011; Klima- og miljødepartementet, 2013). However, most participants are silent about Cartagena, and interviewees noted that this lack of knowledge about Cartagena has made it prone to criticism from Parties with other agendas. This is a strategic choice and a balancing act: "The biggest success is that the C[artagena] D[ialogue] is quiet. It works quietly at the COPs" (Table 6.1, Interview 3). However, it seems that inviting neighbouring countries to meetings is a way of demystifying Cartagena.

Organising and advancing knowledge refer to how Cartagena participants over the years have organised themselves around the domain of knowledge, and how they seek to advance their new shared knowledge to gain influence in negotiations. This is also a practice of *diplomatic learning*. The usual Cartagena participants (not the ministers) are professionals, state officials with different levels of technical and specific knowledge of the different UNFCCC negotiation issues. Practices reflect this shared and mixed background of technical and diplomatic knowledge and experiences. Cartagena participants have organised knowledge in lead teams responsible for specific issue areas, which follow the main UNFCCC negotiation issues. Leading up to Paris, these were mitigation, adaptation, finance, technology transfer, national reporting (including monitoring, reporting, verification, MRV), ADP workstreams 1 and 2, and loss and damage. Lead teams consist of two to three delegations (usually one official from each) spanning North and South. Leads are responsible for developing and drafting notes for discussion at the meetings. This leadership role is important, since this practice is the initial process of bridge-building of specific issues. Failure to deliver useful bridge-building notes risks undermining the sense of common cause and identity.

Cartagena does not negotiate text through consensus. Notes from leads are discussed and modified, but not agreed upon. Parties can use them as they see fit in their own negotiation strategies, knowing well how other Parties stand on the issues. Cartagena participants have several options for using these notes in other negotiation practices: they can use them in developing proposals within their formal political groups, or they can put forward their own interpretation of the notes in the formal UNFCCC negotiations e.g. as non-papers, submissions, or statements. When notes developed within Cartagena find their way into formal negotiations, there is a strong likelihood of Cartagena participants supporting each other's proposals (without referring to Cartagena). This is difficult to observe and prove as an outsider, but I have observed a developing country put forward a rather long text in a formal COP negotiation, which gained wide support from Cartagena participants. Through triangulation with different sources (interviews, negotiation texts, observations), I am confident of the connection. In addition, the UNFCCC secretariat and the co-chairs of an agenda item have used Cartagena as a sounding board for new proposals. At COP16 and COP17, Cartagena played a crucial role as a way of testing and developing compromises, a balanced outcome, and Cartagena proved itself to be a constructive partner in negotiations.

Diplomatic learning also includes a training component for more junior negotiators, or even ministers who are new to climate negotiations. For developed country participants, several mentioned that meetings in developing countries often involve site visits and presentations about local climate policies, activities, and challenges, which for some can be a revelation. As examples given above also illustrate, *diplomatic learning* goes beyond Cartagena when the carefully crafted messages and bridging efforts are fruitful. One interviewee mentioned that the idea of Intended Nationally Determined Contributions had actually matured in Cartagena (Table 6.1, Interview 23). Others mentioned that the experiences and learning in Cartagena had been instrumental in forming AILAC and the High Ambition Coalition (HAC) (Table 6.1, Interviews 21, 22).

The challenges ahead for Cartagena

Communities of practice experience challenges most obviously if participants fail to connect and develop trust, if the domain of knowledge does not arouse passion, and/or if the shared practices do not develop, but remain stagnant (Wenger et al.,

2002, p. 140). This is not the case for Cartagena, which seems to be a strong community of practice. However, other challenges exist:

> [C]ommunity disorders are frequently an extreme version of a community's strength. The very qualities that make a community an ideal structure for learning – a shared perspective on a domain, trust, a communal identity, long-standing relationships, an established practice – are the same qualities that can hold it hostage to its history and its achievements. The community can become an ideal structure for avoiding learning.
>
> (Wenger et al., 2002, p. 141)

Reflecting on Cartagena's strengths and challenges mentioned in interviews, I will further discuss Cartagena's challenges ahead. Strengths are informal dialogue instead of formal negotiation, bridge-building instead of trench-digging, strong personal relationships that build trust, and diplomatic learning across diverse experiences. The risk of relying too much on informal dialogue is that Cartagena never moves beyond talk. We might understand the establishment of AILAC, with many Parties from Cartagena, as moving further beyond informal dialogue to be able to make specific submissions and statements and to better counterbalance the LMDC (Table 6.1, Interviews 11, 21, 22). As documented elsewhere, however, Cartagena and AILAC are actually complementing each other, strengthening the same meta-narrative in negotiations about bridge-building and sharing responsibilities to act on climate change (Blaxekjær & Nielsen, 2015; Watts, 2020). External dynamics can also challenge an informal dialogue. Especially at COP19, bridge-building in Cartagena and elsewhere was hampered as developed countries failed to deliver on finance promises (Oxfam, 2013; ENB, 2013, p. 28; observations). However, Cartagena keeps talking and reaching out, testament to its strong community.

Cartagena's informal nature leads to structural challenges. Cartagena has no formal recognition within UNFCCC, so at COP18 in Qatar and COP19 in Poland, with host governments connected with the fossil fuel economy, it was easier for the COP Presidencies to exclude Cartagena, as they were not obliged to include and utilise Cartagena's expertise and capabilities. Cartagena can initiate meetings with other UNFCCC actors, but Cartagena cannot be sure of being invited and listened to. If Cartagena gets more recognition in the wider field for its expertise and capabilities, then COP Presidencies or co-chairs more or less have to involve Cartagena to secure legitimacy of the process. Cartagena participants have supported each other's proposals to host a COP, and at COP20 in Peru and COP21 in France, both Cartagena participants, Cartagena again played a significant role as a sounding board. Second, since smaller delegations have difficulties with allocating time and resources for actual participation and contributions, and other formal groups and negotiations have to be prioritised, Cartagena could benefit from a stronger focus on diplomatic learning projects with smaller developing countries (see also Castro & Klöck, 2020)

Cartagena's emphasis on personal relationships as a basis for trust and diplomatic learning leads to actor-specific challenges. Although many officials spend many years in the same delegations and have followed UNFCCC negotiations for a long time, replacements in national bureaucracies do happen, and this of course changes the personal composition of Cartagena. All communities must find ways of dealing with this type of change, and interviewees suggest that Cartagena is able to transfer the culture of trust. Newcomers are met with a greater level of trust than before Cartagena was established. Second, people leave with their specialised knowledge and capabilities e.g. if they were part of a lead team. These people are difficult to replace and Cartagena would need to have some measure of documentation and accumulation of learning so as not to lose valuable knowledge. Usually, national delegations have strong measures of documentation, and although not mentioned by interviewees, I expect the Cartagena secretariat to also document shared practices. The real strength of a community of practice is that knowledge, experience, and learning are anchored in practices, not primarily in individuals' minds.

Conclusion

The Cartagena Dialogue for Progressive Action is a significant governance innovation in climate negotiations. After the failure of COP15, where many Parties experienced being left out, Cartagena has become that community of practice, where Parties supporting an action-oriented approach can engage with each other and find common ground based on in-depth knowledge of other Parties' positions and reasoning. Personal ties and strong commitments from Parties made Cartagena able to not only bring trust back into the UNFCCC at a critical moment, but this community of practice was also able to deliver specific suggestions and fine-tuned messages on key issues which helped deliver the Cancún Agreements, the Durban Platform, and the Paris Agreement. Cartagena is an honest and open space to explore ideas without being tied (too much) to national and group positions. It is a community that thrives on boundary spanning as a defining practice whereby necessary face-to-face dialogue and knowledge sharing between North and South and between political groups actually take place. As such, Cartagena creates the basis for trust, and diplomatic learning within and outside Cartagena. The UNFCCC secretariat and especially COP16, COP17, COP20, and COP21 Presidencies were able to use Cartagena as a sounding-board for difficult issues. The trust, experiences, and diplomatic learning in Cartagena also led to the establishment of coalitions that are able to be more vocal in negotiations, namely AILAC and HAC.

Despite Cartagena's important role and contributions in climate negotiations, it has received very little academic attention. In this chapter, I have sought to shed some light on how Cartagena as a community of practice contributes to moving the negotiations forward. More research is needed, first to analyse the years after the Paris Agreement, and second to further develop the theoretical concepts of diplomatic learning and trust in climate negotiations.

Appendix

Table 6.1 List of interviews with Cartagena participants

No.	Interviewees	Interview date
1	Negotiator from developing country delegation. Occasional participant in Cartagena.	26 November 2012
2	Senior negotiator from developing country delegation. Participant in Cartagena since establishment.	3 December 2012
3	Senior negotiator from developing country delegation. Participant in Cartagena since establishment.	3 December 2012
4	Negotiator from developed country delegation. Participant in Cartagena since 2011.	13 December 2012
5	Senior negotiator from developed country delegation. Participant in Cartagena since establishment.	13 December 2012
6	Negotiator from developed country delegation. Participant in Cartagena since 2011.	23 January 2013
7	Minister-level politician from developing country delegation. Participant in Cartagena since establishment.	16 April 2013
8	Minister-level politician from developed country delegation. Participant in Cartagena since COP17.	12 June 2013
9	Senior negotiator from developed country delegation. Participant in Cartagena since establishment.	16 November 2013
10	Senior negotiator from developed country delegation. Participant in Cartagena since establishment.	16 November 2013
11	Negotiator from developing country delegation. Participant in Cartagena since COP18.	18 November 2013
12	Negotiator from developed country delegation. Participant in Cartagena in 2010 only.	18 November 2013
13	Senior negotiator from developed country delegation. Participant in Cartagena since establishment.	18 December 2013
14	Senior expert on developing country delegation. Participant in Cartagena since establishment until COP17.	19 December 2013
15	Senior negotiator from developed country delegation. Secretariat function in Cartagena 2010–2014.	14 February 2014
16	Senior negotiator from developing country delegation. Participant in Cartagena since establishment.	3 December 2014
17	Senior negotiator from developed country delegation. Participant in Cartagena since establishment.	6 December 2014
18	Senior negotiator from developing country delegation. Participant in Cartagena since establishment.	6 December 2014
19	Senior negotiator from developed country delegation. Participant in Cartagena since establishment.	10 December 2014
20	Senior negotiator from developing country delegation. Participant in Cartagena since establishment.	12 December 2014
21	Former senior negotiator from developing country. Participant in Cartagena since establishment.	3–12 December 2015
22	Senior negotiator from developing country delegation. Participant in Cartagena since establishment.	7 December 2015
23	Senior expert on developing country delegation. Participant in Cartagena since establishment.	10 December 2015

Notes

1 This chapter builds on my PhD dissertation (Blaxekjær, 2015), an article on new political groups in UNFCCC (Blaxekjær & Nielsen, 2015), and a chapter on environmental diplomacy (Blaxekjær, 2016).
2 See also UNFCCC (2014a, c) for an official account.
3 A similar definition is given by sources closer to Cartagena (e.g. IISD, 2014; Lynas, 2011).
4 Neither Adler and Pouliot (2011) nor Wenger et al. (2002) treat trust as a theoretical concept. It is beyond the scope of this chapter to discuss trust from a theoretical perspective, and I also apply trust as a practical notion, but see e.g. the edited volume *Whom Can We Trust?* (Cook, Levi, & Hardin, 2009).
5 It is beyond the scope of this chapter to discuss boundary spanning as a theoretical concept, although it is mentioned by Wenger et al. (2002); see also Williams (2002).

References

ABC. (2014). Marshall Islands minister unsure of Australia's stance on climate change. *Australia Network News*.

Adler, E. (2008). The spread of security communities: Communities of practice, self-restraint, and NATO's post-cold war transformation. *European Journal of International Relations, 14*, 195–230.

Adler, E., & Barnett, M. (1998). *A Framework for the Study of Security Communities*. Cambridge: Cambridge University Press.

Adler, E., & Haas, P. M. (1992). Conclusion: Epistemic communities, world order, and the creation of a reflective research program. *International Organization, 46*(1), 367–390.

Adler, E., & Pouliot, V. (Eds.). (2011). *International Practices*. Cambridge: Cambridge University Press.

Adler-Nissen, R. (2014). Stigma management in international relations: Transgressive identities, norms, and order in international society. *International Organization, 68*(1), 143–176.

Araya, M. (2011). The Cartagena dialogue: A Sui Generis Alliance in the climate negotiations. *Intercambio Climático*. Retrieved from http://intercambioclimatico.com /en/2011/02/02/the-cartagena-dialogue-a-sui-generis-alliance-in-the-climate-negotiat ions/.

Blaxekjær, L. Ø. (2015). *Transscalar governance of climate change: An engaged scholarship approach* (PhD thesis). University of Copenhagen, Copenhagen.

Blaxekjær, L. Ø. (2016). New Practices and Narratives of Environmental Diplomacy. In G. Sosa-Nunez & E. Atkins (Eds.), *Environment, Climate Change and International Relations* (pp. 127–143). Bristol: E-IR Publishing.

Blaxekjær, L. Ø., Lahn, B., Nielsen, T. D., Green-Weiskel, L., & Fang, F. (2020). The Narrative Position of the Like-Minded Developing Countries in Global Climate Negotiations. In C. Klöck, P. Castro, F. Weiler, & L. Ø. Blaxekjær (Eds.), *Coalitions in the Climate Negotiations*. Abingdon: Routledge.

Blaxekjær, L. Ø., & Nielsen, T. D. (2015). Mapping the narrative positions of new political groups under the UNFCCC. *Climate Policy, 15*(6), 751–766.

C2ES. (2015). *Toward 2015: An International Climate Dialogue*.

Casey-Lefkowitz, S. (2010). New International Agreement to Fight Climate Change Found Spirit for Consensus in Cartagena Dialogue Countries. In *Switchboard: Natural Resources Defense Council Staff Blog*.

Castro, P., & Klöck, C. (2020). Fragmentation in the Climate Change Negotiations: Taking Stock of the Evolving Coalition Dynamics. In C. Klöck, P. Castro, F. Weiler, & L. Ø. Blaxekjær (Eds.), *Coalitions in the Climate Change Negotiations*. Abingdon: Routledge.

Christoff, P., & Eckersley, R. (2013). *Globalization of the Environment*. Plymouth: Rowman & Littlefield Publishing Group.

Commonwealth of Australia. (2011). Annual Report 2010–2011. Canberra: Department of Climate Change and Energy Efficiency.

Cook, K. S., Levi, M., & Hardin, R. (2009). *Whom Can We Trust?: How Groups, Networks, and Institutions Make Trust Possible*. New York: Russell Sage Foundation. https://ww w.russellsage.org/publications/whom-can-we-trust

Cox, A. (2005). What are communities of practice? A comparative review of four seminal works. *Journal of Information Science, 31*(6), 527–540.

Depledge, J., & Yamin, F. (2009). The Global Climate-Change Regime: A Defence. In D. Helm & C. Hepburn (Eds.), *The Economics and Politics of Climate Change* (pp. 433–453). Oxford: Oxford University Press.

Dimitrov, R. S. (2010). Inside Copenhagen: The state of climate governance. *Global Environmental Politics, 10*(2), 18–24.

Elliott, L. (2013). Climate Diplomacy. In A. F. Cooper, J. Heine, & R. Thakur (Eds.), *The Oxford Handbook of Modern Diplomacy* (pp. 840–856). London: Oxford University Press.

ENB. (2009). Summary of the Copenhagen climate change conference: 7–19 December. *Earth Negotiations Bulletin, 12*(459), 1–30.

ENB. (2010). UN climate change conference in Cancún: 29 November–10 December 2010. *Earth Negotiations Bulletin, 12*(487), 1–2.

ENB. (2013). Summary of the Warsaw climate change conference: 11–23 November. *Earth Negotiations Bulletin, 12*(594), 1–32.

Haas, P. M. (1992). Introduction: Epistemic communities and international policy coordination. *International Organization, 46*(1), 1–35.

Hajer, M. A. (1995). *The Politics of Environmental Discourse: Ecological Modernization and the Policy Process*. London: Oxford University Press.

Herold, A., Cames, M., & Cook, V. (2011). The Development of Climate Negotiations in View of Durban (COP 17). In *Directorate General for Internal Policies Policy Department A: Economic And Scientific Policy, PE 464.459*. Brussels: European Parliament.

Herold, A., Cames, M., Cook, V., & Emele, L. (2012). The Development of Climate Negotiations in View of Doha (COP 18). In *Directorate General for Internal Policies Policy Department A: Economic And Scientific Policy, PE 492.458*. Brussels: European Parliament.

Herold, A., Cames, M., Siemons, A., Emele, L., & Cook, V. (2013). The Development of Climate Negotiations in View of Warsaw (COP 19). In *Directorate General for Internal Policies Policy Department A: Economic And Scientific Policy, PE 507.493*. Brussels: European Parliament.

IISD. (2012). News. Cartagena dialogue discusses expectations ahead of Bangkok and Doha. Retrieved from http://climate-l.iisd.org/news/cartagena-dialogue-discusses-expectations-ahead-of-bangkok-and-doha/.

IISD. (2014). News. Cartagena dialogue to accelerate preparations for post-2020 targets. Retrieved from http://climate-l.iisd.org/news/cartagena-dialogue-to-accelerate-preparations-for-post-2020-targets/.

Independent Diplomat. (2015). RMI/Climate change. Retrieved from http://www.inde pendentdiplomat.org/our-work/rmi-/-climate-change.

Klima- og miljødepartementet. (2013). *Videoblogg fra Warszawa - dag 4* [Videoblog from Warsaw – day 4]. Oslo: Department of Climate Change, Ministry of Climate and Environment, Government of Norway.

Klöck, C., Weiler, F., Castro, P., & Blaxekjær, L. Ø. (2020). Introduction. In C. Klöck, P. Castro, F. Weiler, & L. Ø. Blaxekjær (Eds.), *Coalitions in the Climate Negotiations.* Abingdon: Routledge.

Lynas, M. (2011). Thirty 'Cartagena Dialogue' countries work to bridge Kyoto gap. Retrieved from http://www.marklynas.org/2011/03/thirty-cartagena-dialogue-countr ies-work-to-bridge-kyoto-gap/.

Maldives. (2010a). Address by President Nasheed at the Second Meeting of the Cartagena Group/Dialogue for Progressive Action, 17 July 2010, Ref: PRO/RMN/2010/112.

Maldives. (2010b). Maldives Hosts Meeting for Forward Looking Climate Change Nations, 17 July 2010, Ref: 2010–526.

Marks, K. (2014, July 17). Australia scraps carbon tax: Tony Abbott makes his country a 'global pariah' after legislation is passed by Senate. *The Independent.* Retrieved from http://www.independent.co.uk/environment/climate-change/australia-scraps-carbon-tax-tony-abbott-makes-his-country-a-global-pariah-after-legislation-is-passed-by-sena te-9613291.html.

Meilstrup, P. (2010). The Runaway Summit: The Background Story of the Danish Presidency of COP15, the UN Climate Change Conference. In *Danish Foreign Policy Yearbook* 2010 (pp. 113–135). Copenhagen: Danish Institute for International Studies (DIIS).

Monheim, K. (2015). *How Effective Negotiation Management Promotes Multilateral Cooperation. The Power of Process in Climate, Trade, and Biosafety Negotiations.* London and New York: Routledge.

Neumann, I. B. (2002). Returning practice to the linguistic turn: The case of diplomacy. *Millennium: Journal of International Studies, 31*(3), 627–651.

Oxfam (2013, November 11). Poor countries left in the dark on climate finance: COP19. [Press release]. Retrieved from https://www.oxfam.org/en/press-releases/poor-count ries-left-dark-climate-finance-cop19

Pouliot, V. (2008). The logic of practicality: A theory of practice of security communities. *International Organization, 62*(2), 257–288.

Pouliot, V. (2010). The materials of practice: Nuclear warheads, rhetorical commonplaces and committee meetings in Russian–Atlantic relations. *Cooperation and Conflict, 45*(3), 294–311.

Pouliot, V. (2013). Methodology. In R. Adler-Nissen (Ed.), *Bourdieu in International Relations: Rethinking Key Concepts in IR* (pp. 45–58). London and New York: Routledge.

Readfearn, G. (2014, June 16). What does Australian prime minister Tony Abbott really think about climate change? *The Guardian, Planet Oz.* Retrieved from http://www.theg uardian.com/environment/planet-oz/2014/jun/16/what-does-australian-prime-minister -tony-abbott-really-think-about-climate-change.

RTCC. (2014). *Marshall Islands President: Future Will Be 'Like Living in a War Zone'.* Climate Home News: https://www.climatechangenews.com/2014/04/01/marshall-i slands-president-future-will-be-like-living-in-a-war-zone/

Sebenius, J. K. (1992). Challenging conventional explanations of international cooperation: Negotiation analysis and the case of epistemic communities. *International Organization, 46*(1), 323–365.

UNFCCC. (2014a). *Durban: Towards Full Implementation of the UN Climate Change Convention.*

UNFCCC. (2014b). *Glossary of Climate Change Acronyms.*

UNFCCC. (2014c). *Now, Up to and Beyond 2012: The Bali Road Map.*

UNFCCC. (2015). Paris Agreement. In *Contained in Document FCCC/CP/2015/10/Add.1.*

Van de Ven, A. H. (2007). *Engaged Scholarship: A Guide for Organizational and Social Research.* Oxford: Oxford University Press.

Watts, J. (2020). AILAC and ALBA: Differing Visions of Latin America in Climate Change Negotiations. In C. Klöck, P. Castro, F. Weiler, & L. Ø. Blaxekjær (Eds.), *Coalitions in the Climate Negotiations.* Abingdon: Routledge.

Wenger, E., McDermott, R., & Snyder, W. M. (2002). *A Guide to Managing Knowledge: Cultivating Communities of Practice.* Harvard: Harvard Business School Press.

Williams, P. (2002). The competent boundary spanner. *Public Administration, 80*(1), 103–124.

Willis, A. (2010). UN method hangs in the balance as climate talks begin. EUobserver: https://euobserver.com/environment/31370

Yamin, F., & Depledge, J. (2004). *The International Climate Change Regime: A Guide to Rules, Institutions and Procedures.* Cambridge: Cambridge University Press.

7 The narrative position of the Like-Minded Developing Countries in global climate negotiations

Lau Øfjord Blaxekjær, Bård Lahn, Tobias Dan Nielsen, Lucia Green-Weiskel, and Fang Fang

Introduction[1]

From the inception of the UNFCCC, developing countries' participation in climate change negotiations have been fundamentally shaped by their alliance as a "Southern collective" (Najam, 2005) through the Group of 77 and China (hereafter G77) (Chan, 2020). Now comprising 134 countries, the G77 is marked by large and growing heterogeneity in terms of members' prosperity and political power, greenhouse gas emissions, and vulnerability to climate change. This heterogeneity has given rise to many claims about an inevitable fragmentation or "split" within the G77 (Kasa, Gullberg, & Heggelund, 2008; Roberts, 2011; observations at COP17–21). In later years, coalitions among developing countries in UNFCCC negotiations proliferated (Blaxekjær & Nielsen, 2015). Throughout the history of multilateral climate change negotiations, most developing countries have pursued a strategy of multiple, overlapping alliances, i.e., they have combined membership in the G77 with membership in regional coalitions such as the African Group of Negotiators (AGN), or more climate-specific, interest-based coalitions such as the Organization of Petroleum Exporting Countries (OPEC) or the Alliance of Small Island States (AOSIS) (Castro & Klöck, 2020).

One interest-based coalition that emerged within the G77 in the negotiations leading to the 2015 Paris Agreement is the Like-Minded Developing Countries (LMDC). The LMDC was very vocal and important in the negotiations leading to the Paris Agreement, where the LMDC sought to uphold the differentiation of Annex I vs non-Annex I countries, based on the historical responsibility of the former, equity, and the need to maintain the Convention's core principle of "common but differentiated responsibilities and respective capabilities" (CBDR-RC). In the end, the LMDC and others could not include their core demand for differential treatment of developed and developing countries in the Paris Agreement. The Paris Agreement contains binding commitments for *all* countries (namely, regular submission of nationally determined contributions, NDCs), but does not contain quantified emission reductions for any country (as opposed to the Kyoto Protocol). In this context, we note that the Convention and Kyoto Protocol refer to CBDR-RC, whereas the Paris Agreement added "in light of national circumstances", leading to a discrepancy between the Convention and the Paris Agreement. Nevertheless, the LMDC has been influential in the UNFCCC process and its participation and position are key to understanding the negotiation dynamics, around the 2015 Paris Agreement and beyond. The emergence of the LMDC may be understood on the one hand as a sign of increasing fragmentation and divergence of interests and positions within the G77 – as an expression of change – *and*, on the other hand, LMDC may be understood as an expression of the identity and historical experiences that contributes to a continued, and perhaps surprising, cohesion among Southern countries in increasingly different economic and political circumstances.

In this chapter, we therefore examine the LMDC in more detail, with a focus on the period leading up to the Paris Agreement (2012–2015). Empirically, we collect information from official LMDC statements, from observations at UNFCCC COPs (2011–2015), and in particular from interviews with delegates and experts. Theoretically, we draw on a narrative approach to International Relations and the work by Blaxekjær and Nielsen (2015) on narrative positions in the UNFCCC, and seek to identify LMDC self-perceptions of identity, problems, and solutions to further detail the LMDC narrative position in the UNFCCC.

Our analysis thus yields four central characteristics that make up the LMDC narrative position: first, LMDC perceives itself as an integral part of the G77 and as the "true" voice of developing countries. Second, LMDC is the guardian of the Convention and its principles, most importantly, CBDR-RC and equity. Third, developing countries are the victims of climate change, not the culprits. Historically, they have contributed very little to global climate change. Fourth, even if developed countries must take the lead, developing countries – including the LMDC as coalition and individual countries – are not "blockers", but actively contribute to global climate action.

In the following, we first present our theoretical approach based on the concepts of narratives and narrative positions, as well as explaining our methodological approach and use of submissions, observations, and interviews. We trace the emergence and membership of the LMDC, before turning to the results of our analysis. Here, we discuss in more detail the four core characteristics of the LMDC narrative position – the LMDC as true representative of the G77; the LMDC as guardian of the Convention; developing countries as victims; and LMDC as not blocking. We then discuss the overall findings and conclude.

Theoretical and methodological approach

The chapter draws on Wagenaar's (2011) narrative approach and Blaxekjær and Nielsen's (2015) narrative analysis of UNFCCC negotiations to identify the dominant narrative of the LMDC and map its position in the UNFCCC negotiations landscape. The basic assumption is that language, through e.g. frames or narratives, profoundly shapes our view of the world and reality, instead of merely being a neutral medium mirroring it (Fischer & Forester, 1993; Wagenaar, 2011).

Meaning is not given by a phenomenon in itself, but is established through inter-subjective, linguistic practices, such as narratives (Demeritt, 2001; Nielsen, 2014; Pettenger, 2007).

We define narratives as means by which actors make sense of the world, a "mode of knowing", "providing distinctive ways of ordering experience, of constructing reality", such as an organisation's origin and identity (Wagenaar, 2011, p. 209). This way of ordering is further understood in a simple but recognised template given by Aristotle, where the narrative brings

> unity of action ... from the linear sequence that runs from a beginning, in which the protagonist of the story is faced with a challenge or puzzle, via a middle section in which the events develop, to a final section in which the initial challenge is met or puzzle solved.
>
> (Wagenaar, 2011, p. 210)

We acknowledge that puzzles such as political negotiations on climate change might never be solved, but the point is rather that narratives present versions of future solutions. Narratives also serve a "larger purpose of allowing humans to affirm and reaffirm identities" (Cobley, 2001, p. 222), which is why it is relevant to apply a narrative analysis to new negotiation coalitions in the UNFCCC. Narratives then allow us to understand the arguments (strategic rhetoric) of these coalitions and examine how these arguments are also expressed and reaffirmed in narrative form as identity through the very organisation of the coalition, thus also reinforcing an identity-based imperative for certain action.

Narrative can mean many things, as it has also gained popularity in everyday language and political analyses, and is in many cases interchangeable with rhetoric or story. However, in a stricter academic sense, narratives are understood as the way stories are told – not the stories themselves and not the mere rhetoric or arguments used (Bevir, 2005; Cobley, 2001). A story can be told in many different ways. Narratives bring focus to characters, their relations, progression through time; past, present, and future; problems and solutions. In other words, a narrative is a way of making a coherent connection between identities and action. Strong narratives build on a series of techniques and practices like using familiar templates from fiction, history, culture, and religion often involving questions of justice and appropriate behaviour and/or using enactment and story-telling. In other words, narratives have to be iterated, reiterated, and remembered across time by (many) actors to work effectively as identity formation. Following Wagenaar (2011, p. 218), we focus on "the work that stories do in a particular political or administrative context in order to bring out the story's impact on policy making". Thus, narratives also imply the taking of certain actions (Wagenaar, 2011, p. 215) or, we might add, narratives imply a certain action space. The emergence of new narratives or coalitions shape understandings of what the goals of the UNFCCC negotiations ought to be, and how they should or could be reached.

The concept of *narrative position* (Blaxekjær & Nielsen, 2015) allows us to combine argumentative and organisational practices in the same analysis of a political landscape. A narrative position is a way of situating a narrative (identity, action, etc.) in relation to other narratives of other coalitions and actors in an organisational perspective. The UNFCCC negotiation coalitions are rewriting and reiterating the political landscape of climate change governance. In the UNFCCC negotiations, the LMDC is just one actor telling stories about the global South (the G77), the North (Annex I countries), and how these characters are faring in climate negotiations. Other actors (e.g. the USA, EU, or AILAC) tell the same story, but emphasise different aspects: with a different beginning, middle, and possible endings; with different characters playing different roles (e.g. hero and villain or the innocent victim and perpetrator). Blaxekjær and Nielsen (2015) identified seven new coalitions established between 2009 and 2012 leading up to the Paris Agreement,[2] and visualised their narrative positions according to the main organising principles of the UNFCCC negotiations: Annex I/non-Annex I differentiation and a narrative dimension of "bridge-building" or "upholding the North–South divide" in relation to CBDR-RC:

> LMDC take[s] up a narrative position towards the area where non-Annex I membership aligns with differentiated action. This meta-narrative states that all negotiations shall be under the existing UNFCCC principles on CBDR/ RC and equity, and developed countries must take the lead and raise their ambitions, at a minimum implementing that to which they have agreed. Developing countries are the victim of developed countries' historic emissions. Their responsibility is to pull their people out of poverty and must therefore receive financial and technical support for their differentiated responsibility.
>
> (Blaxekjær & Nielsen, 2015, p. 11)

This chapter is a continuation of this research and examines the LMDC in more detail.

Empirically, we rely on a diversity of sources. Getting direct access to observe and record every LMDC practice and internal working has not been possible for this study, first, because of the time and geographical span of the LMDC, and second, because LMDC meetings and other activities are closed to most observers and media. As with other studies of diplomacy, practices often need to be interpreted from and through other sources. "The rationale is that, even when practices cannot be 'seen', they may be 'talked about' through interviews or 'read' thanks to textual analysis" (Pouliot, 2013, p. 49). We therefore applied a mixed methods approach based on several primary sources (see Tables 7.3, 7.4, and 7.5). We conducted seven interviews with key LMDC representatives from country delegations and unofficial secretariat (Table 7.2, Interviews 1–7), six interviews with observers with LMDC ties (Table 7.2, Interviews 8–13), and seven interviews with experienced negotiators and observers without ties to LMDC (Table 7.2, Interviews 14–20). We made observations specifically on LMDC at COP18, COP19, ADP 2.5 and ADP 2.6 (both Bonn, 2014), COP20, and COP21. We also made observations at COP15, COP16, COP17, ADP 2.2 (Bonn, 2013), and ADP

2.9 (Bonn, 2015), but only indirectly in relation to LMDC. The many observations over six years at several UNFCCC meetings have given us invaluable background knowledge of how UNFCCC negotiations work in practice, helped fine tune our ears to the different narratives, and helped build strong relations with UNFCCC delegates and observers. We also collected and analysed 53 first-hand LMDC sources like statements, submissions, and press releases, with the majority dating from 2013 and 2014.

We argue that this mixed methods approach with a range of sources is one of the best ways of learning about the LMDC and narratives of the climate negotiations, as internal documents and meetings are not open to outside researchers. We gained access to interviewees through networks and/or direct contact at UNFCCC events. We used semi-structured interview guides focusing on getting interviewees to describe the LMDC's history and purpose, how they see problems and solutions, as well as the way the LMDC functions throughout the year and during UNFCCC sessions. It is our understanding that the interviewees' answers are trustworthy and valid. When issues were too sensitive to talk about, interviewees said so and refrained from answering. Interviewees emphasised different aspects relating to their personal experiences, which indicates that answers were not rehearsed and officially sanctioned. No discrepancies were found when triangulating with other sources, although there is some ambiguity in the LMDC narrative. Through interviews with LMDC representatives and other observers of UNFCCC negotiations, we can identify narratives of and by LMDC, and then triangulate with other sources such as LMDC statements, submissions, and press releases, as well as other publicly available sources like interviews with negotiators in newspapers. At least two members of the research team transcribed, read, and coded the interviews and other written sources. We coded to further detail the LMDC narrative position of LMDC origin, purpose, solutions, and problems in the UNFCCC negotiations as identified by Blaxekjær and Nielsen (2015).

The LMDC history and membership

Before turning to our narrative analysis and the four key characteristics of the LMDC's narrative position, we describe the LMDC's establishment and organisation, and classify the coalition based on the criteria in Castro and Klöck (2020): geographic and thematic scope, membership, and level of formality.

The first formal submissions and statements delivered by the LMDC appeared in May 2012 (LMDC, 2012a–c).[3] This followed a meeting among some LMDC countries in Geneva in March 2012, as well as preparatory, bilateral talks among key countries (Table 7.2, Interview 2). The Geneva-based intergovernmental policy research think-tank, the South Centre, was instrumental in setting up the first meeting, and has since provided policy support according to its mandate and facilitated meetings when asked to do so by member countries (Table 7.2, Interviews 2, 4, and 11). The LMDC was established, to some degree, as a response to the outcome of COP17 in Durban a few months earlier. The adoption of the Durban Platform and subsequent establishment of the Ad-Hoc Working Group on the

Durban Platform (ADP) as a negotiating track for the post-2020 agreement was widely seen as a setback for the positions of key LMDC countries such as China, India, the Philippines, Malaysia, Saudi Arabia, Venezuela, and Ecuador. The mandate for the Durban Platform made no explicit reference to the principles of CBDR-RC and equity, instead mandating the negotiation of an agreement "applicable to all", placing the issue of differentiation front and centre in the negotiations (Table 7.2, Interview 1).

Many LMDC interviewees also emphasised the long historical roots of the personal relationships and the political and strategic thinking behind the LMDC. Negotiators of key LMDC countries have worked closely together for many years in the UNFCCC negotiations (Table 7.2, Interviews 1, 8, 9, 14, 15). Before the formation of the LMDC, many of these negotiators were part of an informal grouping known as "The Elders" within the G77, already coordinating their positions and tactics prior to COP15 in Copenhagen (Table 7.2, Interviews 8, 15). In the words of a negotiator from an LMDC country, the LMDC has already "been there for a while [...] as sort of a dormant giant" (Table 7.2, Interview 1).

Interviewees highlighted that achieving consensus on the LMDC's core issues within the G77 had grown increasingly difficult over the last few years. This was linked to the fact that a few (mainly Latin-American) developing countries started to coordinate their positions more closely with developed countries through the meetings of the Cartagena Dialogue, and with each other through AILAC (Table 7.2, Interviews 1, 3, 9, 16, 17). While representatives of LMDC countries refuted the idea that it was the establishment of the Cartagena Dialogue or AILAC in itself that prompted the formation of the LMDC, it was acknowledged that the changed dynamic within the G77 was part of the reason why the LMDC came together after Durban (Table 7.2, Interview 1).

Further, the LMDC establishment is also tied to the strong reactions against China, India, Brazil, and South Africa from other developing countries after the BASIC group took part in negotiating the Copenhagen Accord document directly with the US during the final hours of COP15. After COP15, important actors within the G77 criticised the BASIC countries for lack of loyalty to the wider Southern collective, because of their role in negotiating an outcome behind the back of other developing countries (Table 7.2, Interviews 2, 8).

The LMDC membership and leadership is not fixed and changes from submission to submission and meeting to meeting (Table 7.2, Interview 2). The country list from the Conference Room Paper by Malaysia (dated 3 June 2014) includes 19 countries[4] as opposed to the 38 countries from May 2012. At the annual LMDC strategy meeting held in India on the 14–15 September 2015, negotiators from 13 countries participated (LMDC 2015d).[5] There is no formal list of members, preparation takes place via email (with no fixed list of recipients), and meetings have few to many participants. We were not able to confirm whether there is an LMDC chair, but there is an internal coordinator from the South Centre (Table 7.2, Interview 2). For external communication, a press release in 2013 states that Yeb Saño,[6] then-head of the Philippines delegation, acted as LMDC spokesperson (LMDC, 2013t). Statements and submissions only occasionally list countries by which it is co-sponsored. On 41 submissions and statements related to the ADP for the period 2012–2014, only seven list co-sponsoring countries, shown in Figure 7.1 below.

Based on participation in meetings and submissions, there seems to be a core group of countries. We consider the countries submitting documents, press releases, and delivering statements on behalf of the LMDC as the core countries (see Appendix for list). These countries are the Philippines (12 times), Venezuela (5), Ecuador (4), India (4), Nicaragua (4), Pakistan (4), Egypt (3), Algeria (2), Bolivia (2), Cuba (2), Malaysia (2), Mali (2), Saudi Arabia (2), Argentina, Dominica, Iran, Jordan, Sudan, and Thailand. Regarding how countries are selected to represent the LMDC, one interviewee said it is usually based on will-ingness and suitability, but sometimes just random selection (Table 7.2, Interview 2). We note that China does not take a more public role in the LMDC; however, China should be considered a core country, as it has hosted meetings and delivers different key input and resources (Table 7.2, Interview 2, 3).

Other countries that have been associated in at least one submission or state-ment are Libya, Paraguay, Syria, Yemen, Bahrain, Comoros, Djibouti, Mauritania, Morocco, Oman, Palestine, Somalia, Tunisia, Ghana, Lebanon, and the Maldives. We consider these to be more peripheral countries – some might even be included in the statement as a misunderstanding e.g. the Maldives. Where China and India are able to connect to a stronger developing country identity in LMDC than in

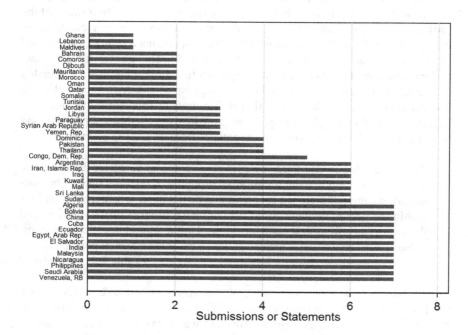

Figure 7.1 Number of explicit mentions in LMDC submissions, statements, and conference room papers in ADP negotiations 2012–2014.

BASIC, where it is much clearer that all members are emerging economies, it should be noticed that Brazil and South Africa have *not* joined LMDC. It is beyond the scope of this chapter to explain this. Some scholars argue that a split in BASIC between China and India on one side, and Brazil and South Africa on the other is becoming visible (Hochstetler & Milkoreit, 2014).

After what appears to have been the initial meeting among some LMDC countries in Geneva in 2012, the coalition has met through annual strategy meetings (in China in 2012 and 2013, Saudi Arabia in 2014, and India in 2015) as well as coordination meetings before and during UNFCCC sessions (TWN, 2012; 2013; LMDC, 2013a, 2015d; Table 7.2, Interviews 1, 2, 4, 11; observations). The LMDC has previously invited other groups, the UNFCCC secretariat, as well as the ADP co-chairs, to such meetings or in conjunction with such meetings. Over time, the organization and coordination of LMDC has grown with policy research and practical support by core countries, the South Centre, Third World Network, and NGOs in LMDC countries (Table 7.2, Interviews 2, 4, 10; observations). China, Saudi Arabia, India, and the Philippines in particular seem to be instrumental in organising the LMDC, providing technical support, including hosting LMDC meetings, preparing negotiation documents and expert advices, and leading external communication, as well as providing financial support to negotiators from smaller LMDC countries to attend LMDC meetings and pre-meetings (Table 7.2, Interview 1, 2, 4). For their annual strategy meetings, the host country is responsible for preparing the agenda and facilitates the meeting (Table 7.2, Interview 2, 4). Interviewees, however, stress the informal and fluid status of the LMDC, consciously not using the word "group" to describe their cooperation. Instead, they refer to themselves with the more open-ended term "Like-Minded Developing Countries" (Table 7.2, Interviews 1, 9). This allows for what several interviewees described as a fluid approach to coordination and membership (Table 7.2, Interviews 1, 8, 15). While a core group of countries is central to the coordination of the LMDC and participates in most of its submissions and joint interventions, other countries may be on the fringes, sometimes supporting and other times not. As explained by one negotiator from an LMDC country:

> the membership is also fluid. You see some constant countries there, but the like-minded group remains an informal grouping. It does not operate by consensus, meaning if there are countries that do not agree with certain positions within the *group*, quote/unquote, all they have to do is just step aside, and the group can still push specific positions. Unlike formal negotiating groups, [where] until consensus is arrived at, you cannot have formal positions.
>
> (Table 7.2, Interview 1)

To summarise, and in the framework of this volume (Castro & Klöck, 2020), the LMDC is a global issue-specific coalition, as members clearly have their roots in the climate change negotiations, and was formed at a specific point in the negotiations around common objectives and values. The LMDC is an informal coalition, although it is well coordinated, with regular meetings and many co-sponsored submissions and statements. We now turn to the narrative position of the LMDC.

Results: the narrative position of the LMDC

LMDC as the true representative of G77

All sources confirm that the LMDC's self-understanding and self-projection is couched in references to the global South – particularly to the G77. The first statements in May 2012 and the LMDC press release, published just after its first annual meeting in China, "stressed that this grouping is part of and is anchored firmly in the G77 & China", and that LMDC "agreed to continue to work together to strengthen the unity of G77 & China and play a constructive and meaningful role in the negotiations" (TWN, 2012). One policy advisor on an LMDC country delegation described LMDC as "a formally organised coalition under the G77. It is a sub-group" (Table 7.2, Interview 3).

This representation of LMDC as a group that is formally under G77 and based on common interests constructs the LMDC as a natural and legitimate developing country community, thereby making it politically stronger internally and externally. The emphasis on its developing country identity and solidarity counters descriptions of the LMDC as blockers (see below) or hardliners (Table 7.2, Interview 17).

By outside observers, the LMDC has been portrayed as

> an alliance between China and India, and oil-producing countries like Saudi Arabia and Venezuela, with a few climate 'radicals' like Bolivia and Ecuador, fast-growing economies like Malaysia and Thailand, and sundry others such as Egypt, Nicaragua, Pakistan and the Philippines thrown in.
>
> (Bidwai, 2014, p. 14)

From this perspective, the LMDC can be understood as a strong negotiation tool for some powerful non-Annex I countries. LMDC is a sort of BASIC (plus), where a powerful group of countries set the agenda, while less powerful countries provide legitimacy (Table 7.2, Interview 8, 17). Indeed, the LMDC appears, to some extent, to have taken over the role of BASIC as a group of powerful developing countries putting pressure on Annex I Parties (Blaxekjær & Nielsen, 2015). From this point of view, the LMDC does not represent the views of developing countries at large, but mainly the hardliners, and is a platform for them to keep a firm rhetoric on core issues (Table 7.2, Interview 17).

On the other hand, LMDC appears to capture existing rhetoric and views held by several members of the G77 (as evident in their statements). This suggests that the LMDC rhetoric is not purely a strategic or reactive imperative, but also a movement that captures a certain core narrative amongst developing countries. Amongst several of the LMDC interviewees, there was a sense that the LMDC did not emerge in response to or in reaction to the Durban Platform, but that it has been there for a while, and can perhaps be more connected to a general dismay with Annex I parties not living up to their commitments (e.g. Table 7.2,

Interviews 1, 4, 8, 10; observations). In voicing and amplifying this dismay, the LMDC also connects to a longer history of North/South power struggles in the international community – a fact explicitly recognised by observers as well as negotiators (Table 7.2, Interviews 1, 15; ENB, 2012a–d; ENB, 2013a–g; ENB, 2014; ENB, 2015a–e; observations).

From this view, the LMDC was formed for a specific purpose – to keep a strong and unified non-Annex I position in the negotiations. This can be seen as a reaction: to increasingly conflicting views on CBDR-RC amongst developing countries after COP15, which made a strong and unified G77 position more difficult; to a concern that the post-2020 framework would bring binding obligations for non-Annex I Parties; and to the non-Annex I critiques of BASIC not being inclusive enough. This view gives the impression of a more reactive than proactive group (e.g. Table 7.2, Interviews 8, 17).

Yet the LMDC emphasises that a common developing country identity based on common interests and solidarity has existed for a long time. Accordingly, interviewees (Table 7.2, Interviews 1–8, 10, 12, 13) downplayed differences within G77 and emphasise instead cooperation and conversations with other G77 sub-groups. As highlighted above, there does exist a sense among developing countries that Annex I countries in many cases have failed to live up to their commitments, as well as a shared sense of history and position at the margins of international politics (cf. Najam, 2005). The persistence of these views among many developing countries has allowed LMDC to position itself to some extent as defenders of developing countries' interests and a true representative of G77 – even if the very emergence of the LMDC points to increasing difficulties of reaching consensus on strong positions within this group.

LMDC as guardian of the Convention

The LMDC sees itself as protecting the Convention from being dismantled. Developing countries have put faith in the process (the Convention), believing that what was originally agreed on would be respected and implemented (Sering, 2013; see Table 7.4). In the LMDC countries' view, developing countries agreed to agree on a post-2020 framework, but not a re-negotiation of the Convention (e.g. LMDC, 2012c; LMDC, 2013x; LMDC, 2014i; Table 7.2, Interview 7). In particular, they agreed to safeguard three core elements of the Convention: the principle of CBDR-RC, historical responsibility, and equity – all of which implies upholding a differentiation of Annex I and non-Annex I parties.

The general concern among LMDC members is that Annex I Parties have not fulfilled their obligations and instead now want more action from non-Annex I Parties (e.g. LMDC, 2014i). As Nitin Desai from the WWF says in an interview: "You can't go down and say look I know I am not doing as much as I claim to do but you better start doing more" (cited in RTCC, 2012). Lack of leadership means that Annex I Parties evade their own responsibilities and shift them towards non-Annex I Parties (cf. interventions by Saudi Arabia, Venezuela, and Malaysia, amongst others, High Level Segment COP19, 20 November 2013).

Typically, LMDC statements warn against replacing, rewriting, and reinterpreting the Convention (e.g. LMDC, 2014i). Hence, removing the annexes or re-interpreting the principle of CBDR-RC is a way for Annex I parties to "backtrack" on their obligations and leave the post-2020 framework toothless, with Parties having little and very vague obligations (Table 7.2, Interviews 6, 7). Changing the rules of the Convention will pre-empt the outcome of the post-2020 framework, deflate the responsibilities of Annex I Parties, and dilute the whole negotiation process as such: "Parties would then do what they want, when they want, how they want", in the words of a high-level LMDC negotiator (Table 7.2, Interview 6; see also LMDC, 2012a). This argument is closely linked with the perception of developing countries as victims, not perpetrators, discussed in the next section.

A key recurring argument of the LMDC is that universality is not the same as uniformity:

> We also believe that the term 'applicable to all Parties' refers to the fact that the outcome of the Durban Platform process will apply to all Parties, similar to how the Convention and the Kyoto Protocol is applicable to all Parties which enter into these treaties. This term does not mean that the outcome must be such that all Parties undertake uniform or similar types and levels of obligations.
>
> (LMDC, 2012a)

In the LMDC narrative, the "re-interpretation" of especially CBDR-RC towards "uniform applicability" in the Durban Platform is a critical break with what they agreed to. Accordingly, there are concerns that moving away from the Convention and the Kyoto Protocol system will significantly reduce the success of a future climate regime:

> What we have seen, especially during the past few years, is a pledge-and-review system that has even worsened after Warsaw, where we are now talking about contributions instead of commitments for all countries. That's dangerous because we need legally binding agreements, legally binding targets, that countries must pursue.
>
> (Table 7.2, Interview 1)

The LMDC argues that altering the Convention risks hampering developing countries' ability to combat climate change. In an interview with E&E News, the Philippines' delegate Yeb Saño argues:

> It's still very relevant, and we think it should be protected. If it is set aside, we have nothing, and what we will have is a hugely disproportionate or biased agreement. It will be something that will prejudice the interests of the most vulnerable nations on Earth because the Convention is something we have not even fully implemented.
>
> (cited in Friedman, 2014)

Developing countries as victims of climate change

The strong objection to binding emission reductions for non-Annex I countries is closely linked with the portrayal of developing countries as *victims* of climate change, having less (or no) historical responsibility for anthropogenic climate change, and not near the capacity of developed countries to mitigate climate change. LMDC members (big and small) regard themselves as standard bearers of fairness and equity, holding developed states – the culprits – to account for their historical responsibility. Developed countries must act first, for developing countries have other pressing concerns and are not responsible for climate change (Table 7.2, Interviews 1, 2, 4).

Furthermore, the LMDC seek to highlight the broad basis that goes well beyond some hardliners and members with large greenhouse gas emissions. Instead, they claim to speak on behalf of the most vulnerable countries to climate change:

> [O]ur populations and economies are also among those that are particularly vulnerable to the adverse effects of climate change. Recent examples are the floods in Argentina, China, Ecuador, Malaysia, Pakistan, Philippines, Sri Lanka, Thailand, and Venezuela; the super-typhoons and flooding in the Philippines; the droughts in Bolivia, Democratic Republic of the Congo, Egypt, India, Mali, Sri Lanka, and Sudan; the extreme rains that affected Algeria, China, Ecuador, El Salvador, and Nicaragua; droughts and melting glaciers in Argentina, Bolivia, and Ecuador; and massive dust storms hitting Iran, Iraq, Kuwait, and Saudi Arabia, among others. These climate change impacts in our countries have killed thousands of people and set back our poverty eradication and sustainable development efforts. These are also the impacts that affect other developing and least-developed countries. In many ways, these impacts make more difficult the situation of small island developing states, least-developed countries, and African countries, given the multiple challenges that they already face.
>
> (LMDC, 2013e)

The LMDC stresses that its member populations, half of which live on less than US$2 a day, face severe development challenges (LMDC, 2013e). The argument is that social and economic developments and poverty eradication are still the first and overriding priorities of many LMDC members (LMDC, 2013w). Iskander (2013; Table 7.4) proposed the UNFCCC establish a list of the most vulnerable countries. LMDC is the guarantor of the interests of the poorest and most vulnerable:

> We also want to put forward a very strong developing country voice, because, after all is said and done, we arrive at a compromise ... So if there is no strong developing country voice, we will arrive at something that is disadvantageous to the poorest people in the world. We don't want that to happen. So that's why we endeavour to constantly build a strong developing country voice in the negotiations on every front.
>
> (Table 7.2, Interview 1)

The LMDC seeks to remind developed countries of their historical responsibility and responsibility to lead the fight against climate change. This is often expressed through criticism of "inadequate" or "extremely disappointing" efforts to reduce emissions and help developing countries through e.g. capacity building and technology transfer (e.g. Table 7.2, Interview 1, 2, 4). The point is that developing countries have yet to see Annex I countries live up to their commitments in terms of both emissions reduction under the Convention and the Kyoto Protocol and also the finance promised to support developing countries in their efforts.

LMDC members are not blockers

Because of their firm stance on historical responsibility, equity, and CBDR-RC, and hence their insistence on the differential treatment of Annex I and non-Annex I parties inherited from the Convention and Kyoto Protocol, the LMDC is often seen by others to "block" negotiation progress (Table 7.2, Interview 17).

In contrast, the LMDC members do not see themselves as "blockers" and react strongly to being called so. As a counter-argument, LMDC portrays the group as being able to come up with solutions (under certain conditions). From the first statements and towards COP21, LMDC has continuously made a strong effort to communicate that LMDC is constructive and ready to reach an agreement.

> [W]e have proactively and constructively put forward textual proposals that reflect the best way through which the Convention's implementation will be fully, effectively, and sustainably enhanced before and after 2020, in which the Convention's principles, provisions, and structure are fully respected and reflected in a way that is in full accordance with the differentiated obligations and commitments of developed and developing countries, and is consistent with Article 4 of the Convention.
>
> (LMDC, 2015c)

Interviews with non-LMDC members show that the LMDC does represent a voice that needs to be there, and that LMDC is not just a blocker (e.g. Table 7.2, Interview 16). In this view, the LMDC represents a forum and a platform that is based on some sensitive issues that strike a chord amongst several non-Annex I countries as well as other stakeholders. This includes frustrations with a lack of leadership by Annex I Parties (see above). Jayanthi Natarajan, Union Environment and Forests Minister of India, illustrates this frustration in an interview with *The Hindu*: "We are by no means a nay-sayer. We only object to any prescriptive policies that are dictated to us by others who are actually not doing anything to combat climate change" (cited in Sethi, 2013).

A recurring argument is that if Annex I countries had lived up to their original commitments (or the 40% reduction by 2020 discussed in Durban), non-Annex I countries would not have had to adapt to climate change and there would be no "emissions gap" (Table 7.2, Interview 6). LMDC argues that further burdens – due to the emissions gap – should not be placed on developing countries. LMDC wants to see developed countries show the way forward and live up to their obligations – not deflect them onto non-Annex I countries – before they can commit to any form of climate action (Table 7.2, Interview 1; LMDC, 2015d).

Nevertheless, the strong objection to binding emission reductions for non-Annex I does not mean that non-Annex I countries and LMDC members are doing nothing to mitigate emissions at home. Quite to the contrary, LMDC often points out that developing countries are indeed already doing much to combat climate change, despite their relatively lower capacities to do so.

> As developing countries face multiple challenges in terms of social and economic development and poverty eradication, the LMDC are undertaking ambitious actions domestically on climate change. ... The LMDC expressed their continued willingness to participate in the negotiations in a constructive, consensus-building, and Party-driven manner to reach an ambitious, comprehensive, equitable, and balanced Paris outcome.
>
> (LMDC, 2015d)

LMDC notes that the total emissions reductions pledged by developing countries linked to the Cancún Agreements are greater in absolute terms than the reductions pledged by developed countries (LMDC, 2013e, f). It should be added, however, that actions on climate change by non-Annex I Parties must be supported by Annex I Parties, according to LMDC (Table 7.2, Interview 1). The contributions by non-Annex I Parties – such as those made at Cancún (COP 16, December 2010) – depend upon the support of Annex I Parties. LMDC demands that non-Annex I parties pay for developing countries' necessary actions. Annex I parties' mitigation and climate-related finance should not be seen as a service, but as repayment of their carbon debt (e.g. Malaysia in TWN, 2015). As such, the LMDC again stresses the leadership and responsibilities of Annex I Parties as the *sine qua non* of a future climate regime. For LMDC countries in the African Group, finance in particular is paramount, as it determines the level of ambition for these countries (Makonga 2013; Table 7.4). Regarding finance, LMDC frequently echoes the language of ALBA members, stressing that there is a need for non-market mechanisms (LMDC, 2013gg), and that the role of the private market in the Green Climate Fund is only subsidiary (LMDC, 2013w).

Furthermore, the large LMDC countries do acknowledge the diversity of developing countries. Even if they insist that a strong differentiation between developed and developing countries be upheld, some LMDC members include a second level of differentiation – if, and only if, the differentiation between Annex I and non-Annex I countries is recognised (Table 7.2, Interview 1). This second level concerns differentiation between developing countries – and/or within a developing country with (great) differences in income distribution. This approach is a based on a narrative of "a distribution of the right to development", or what is sometimes referred to as "the greenhouse development rights approach" (Yeb Saño, Philipppines, cited in Friedman, 2014).

Discussion and conclusion

In the negotiations of a successor agreement to the 1997 Kyoto Protocol, larger developing countries with significant and growing greenhouse gas emissions – notably China, but also other emerging economies – were increasingly under pressure to accept binding emission reduction commitments (Rajamani, 2016; observations). The rigid differentiation between developed (Annex I) and developing (non-Annex I) countries was also increasingly contested within the group of developing countries. While parts of the Global South (such as the LDCs, AOSIS, or AILAC) sought to bridge the North–South divide, others insisted on differential treatment (Blaxekjær & Nielsen, 2015). In particular, LMDC emerged as a coalition around this core question of differentiation. We found that the LMDC has fluid membership, but consists of a core group of 15–20 countries, sometimes joined by another 15–20 countries. Countries like the Philippines, China, India, Saudi Arabia, Malaysia, Venezuela, and Ecuador play leading roles. The LMDC is supported by the intergovernmental organisation, the South Centre, based in Geneva, which has hosted several LMDC meetings and provided policy input. There seems not to be an LMDC chair, but at some point in 2013 at least, Yeb Saño, head of Philippines delegation, was the LMDC spokesperson.

In this chapter, we examined in more detail the narrative position of the LMDC first identified by Blaxekjær and Nielsen (2015), with a focus on the negotiations leading up to the 2015 Paris Agreement. Through mixed methods and different primary and secondary data, notably 20 interviews, observations from COP15 to COP21, and 57 negotiation documents, we identified key characteristics of the LMDC's narrative position. This position focuses on four distinct, but closely related, elements: first, the LMDC considers itself as the "true" representative of the developing world that, second, guards core principles of the Convention – in particular CBDR-RC, equity, and the historical responsibility of developed countries. These three principles imply that developed countries must take the lead and reduce emissions first, and not shift the burden of action to developing countries. These are still poor and developing, and, third, vulnerable victims of the emissions of developed countries. Finally, the insistence on differentiation between developed and developing countries does not make LMDC "blockers"; developing countries are actively reducing emissions at home, and in fact more so than developed countries.

This narrative position echoes the positions of other countries, notably BASIC and ALBA – which is, of course, unsurprising given the core membership of these different groups: India and China are key players in both, BASIC and LMDC, while Venezuela and Bolivia are key players in both, ALBA and LMDC. Similar to BASIC, LMDC promotes itself as unifying and strengthening the position of developing countries in the UNFCCC. The coalition presents itself as being closely aligned with G77, but appears to use a more direct approach in its criticism of the lack of action undertaken by the developed countries. Indeed, LMDC members regard themselves as standard-bearers of fairness and equity, holding rich nations to account for their historical responsibility, and as such, claim to speak for the developing world as a whole, yet the coalition's key demands were only partially met in the Paris Agreement. As opposed to what LMDC demanded, the Paris Agreement finally leaves behind the Kyoto Protocol's rigid differentiation of countries. The Paris Agreement for the first time treats all countries in the same way, requiring both, Annex I *and* non-Annex I countries, to submit NDCs and take climate action at home – although the Paris Agreement does, of course, also acknowledge the diverse capabilities of developing countries by adding "in light of national circumstances" to the principle of CBDR-RC. It also reaffirms the principle of equity, placing it as a starting point for the process of the so-called "global stocktake" in a way that "leaves the door open for a dialogue on equitable burden sharing" between developed and developing countries (Rajamani, 2016, p. 504).

The LMDC narrative position might claim to be representative of developing countries, but the number of countries in the LMDC has not grown over the years. Developing countries are extremely diverse. We began this chapter by observing the increasing fragmentation of the developing world and the emergence of sub-groups of the G77. At first glance, the LMDC adds to this fragmentation. In the run-up to Paris, "Most Annex 1 countries want … a new agreement binding all countries. Rather than reunite the G77, the LMDC could end up further splintering or wrecking it", writes Bidwai (2014, p. 14), adding that "The LMDC group did not manage to influence the Doha or Warsaw climate conference strongly". On the other hand, as Rajamani (2016, p. 504) argues, making the principle of equity part of the basis for the global stocktake can be seen as something of a "negotiating coup" for developing countries, suggesting that the existence of a strong voice of traditional developing country positions may have had some influence on the Paris Agreement. The inclusion of the 1.5°C temperature goal in the Paris Agreement was also seen as victory for the vulnerable developing countries (observations), a theme also included in the LMDC narrative position. If we closely examine the narrative position of the LMDC, we are able to identify the common ground that does exist between the LMDC and the wider group of developing countries, which may help explain how this influence came about. Overall, our research suggests that the emergence and continued organisation and efforts of the LMDC should be understood as a sign of increasing fragmentation and divergence of identities, interests, and positions within the G77, and *simultaneously*, as an expression of the identity and historical experiences that contributes to a continued, and perhaps surprising, cohesion among developing countries in increasingly different economic and political circumstances.

Appendix

Tables 7.1 to 7.5

Table 7.1 Events observed by one of more of the co-authors

Name of event	Place	Time
COP15/CMP5	Copenhagen, Denmark	7–18 December 2009
COP16/CMP6	Cancún, Mexico	29 November–10 December 2010
COP17/CMP7	Durban, South Africa	28 November–9 December 2011
COP18/CMP8	Doha, Qatar	30 November–7 December 2012
ADP2-2	Bonn, Germany	4–13 June 2013
COP19/CMP9	Warsaw, Poland	13–21 November 2013
ADP2-5	Bonn, Germany	4–15 June 2014
ADP2-6	Bonn, Germany	20–25 October 2014
COP20/CMP10	Lima, Peru	1–13 December 2014
ADP2-9	Bonn, Germany	1–11 June 2015
COP21/CMP11	Paris, France	30 November–12 December 2015

Table 7.2 List of interviews

No.		Interview date
Interviewees representing LMDC		
1	High-level negotiator 1	5 June 2014
2	Policy Advisor on LMDC country delegation 1	10 June 2014
3	Policy Advisor on LMDC country delegation 2	6 August 2014
4	Coordinator, IGO	10 June and 23 October 2014
5	High-level negotiator 2	23 October 2014
6	High-level negotiator 3	24 October 2014
7	High-level negotiator 4	25 October 2014
Interviewees with ties to LMDC countries		
8	Representative from NGO network, based in LMDC country	5 June 2014
9	Representative from NGO, based in Europe	5 June 2014
10	Representative from Southern-based NGO with close ties to LMDC negotiators	6 June 2014
11	Representative from a large international NGO, based in Europe, working with developing countries including many LMDCs	10–11 June, 15 August, 21 September, and 7 October 2014
12	Experienced NGO observer, based in LMDC country	28 July 2014
13	Anonymous	24 October 2014
Interviewees without ties to LMDC countries		
14	High-level negotiator, Europe	5 June 2014
15	Representative from NGO, based in non-LMDC African country	6 June 2014
16	Former high-level negotiator, Latin American country	11 June 2014
17	Experienced UNFCCC observer 1	21 September 2014
18	Representative from NGO	23 October 2014
19	High-level negotiator, AILAC	24 October 2014
20	Experienced UNFCCC observer 2	27 October 2014

Table 7.3 List of LMDC written sources

Source	Delivered by	Date
LMDC, 2012a. *Submission on the Ad-Hoc Working Group on the Durban Platform [Joint statement]*	Argentina	17 May 2012
LMDC, 2012b. *Submission on the Ad-Hoc Working Group on the Durban Platform (AWG-DP) [Joint statement]*	Venezuela	22 May 2012
LMDC, 2012c. *Joint statement on the closing plenary session of the ad-hoc working group on the Durban Platform on Enhanced Action*	Philippines	25 May 2012
LMDC, 2013a. *Meeting of the Like-Minded Developing Countries on Climate Change,* Geneva, Switzerland, 27–28 February [Press release]	Ecuador	1 March 2013
LMDC, 2013b. *Implementation of all the elements of decision 1/CP.17, (a) Matters related to paragraphs 2 to 6; Ad-Hoc Working Group on the Durban Platform for Enhanced Action (ADP)* [Submission]	Philippines	13 March 2013
LMDC, 2013c. *Submission on the ADP*	Philippines	13 March 2013
LMDC, 2013d. *Post-2020 Climate Agreement Must be Consistent with 1992 Framework Treaty: "Common but differentiated responsibility", "equity", and developed country ambition key to a successful agreement* [Press release]	Philippines	28 April 2013
LMDC, 2013e. *LMDC Opening Plenary Statement for ADP Bonn Session*	Nicaragua	29 April 2013
LMDC, 2013f. *LMDC Views on Summary of Elements and Management of Work for Workstream 1 in the ADP*	Philippines	29 April 2013
LMDC, 2013g. *LMDC Closing Statement for ADP2-1*	Malaysia	3 May 2013
LMDC, 2013h. *Developing countries doing more to save the climate, developed country leadership and ambition missing* [Press release]	Philippines	3 May 2013
LMDC, 2013i. *LMDC Opening Statement for SBI 38*[7]	Dominica	3 June 2013
LMDC, 2013j. *LMDC Opening Statement for SBSTA 38*	Thailand	3 June 2013
LMDC, 2013k. *LMDC Opening Statement for ADP2-2*	Sudan	4 June 2013
LMDC, 2013l. *LMDC: Successful post 2020 climate agreement depends on developed country leadership before 2020* [Press release]	Philippines	4 June 2013
LMDC, 2013m. *LMDC Statement for ADP Round table on workstream 1: Variety of enhanced actions*	Philippines	6 June 2013

(*Continued*)

Table 7.3 Continued

Source	Delivered by	Date
LMDC, 2013n. *LMDC Statement for ADP Workstream 2: Enhancing pre-2020 ambition*	Venezuela	8 June 2013
LMDC, 2013o. *LMDC Statement for ADP Workstream 1 Informal Roundtable: Finance, Technology, Capacity Building*	Saudi Arabia	8 June 2013
LMDC, 2013p. *LMDC Statement for ADP Informal Plenary*	Cuba	11 June 2013
LMDC, 2013q. *LMDC Statement for ADP Informal Plenary*	The Philippines	12 June 2013
LMDC, 2013r. *Closing Plenary of the Resumed 2nd Session of the ADP*	Pakistan	13 June 2013
LMDC, 2013s. *LMDC Closing Statement for SBSTA 38 delivered by Algeria*	Algeria	14 June 2013
LMDC, 2013t. *LMDC: Implementing core principles of climate treaty key to successful climate talks* [Press release]	Philippines	14 June 2013
LMDC, 2013u. *LMDC Submission on Agriculture, including Fisheries*	Mali	2 September 2013
LMDC, 2013v. *Submission on the costs, benefits, and opportunities for adaptation based on different drivers of climate change impacts, including the relationship between adaptation and mitigation*	Bolivia	12 September 2013
LMDC, 2013w. *Views of the Like-Minded Developing Countries on Climate Change (LMDC) on Workstreams 1 and 2 of the ADP*	Ecuador and El Salvador	24 September 2013
LMDC, 2013x. *19th session of the UNFCCC Conference of the Parties Opening Statement by the Like-Minded Developing Countries in Climate Change (LMDC)*	Nicaragua	11 November 2013
LMDC, 2013y. *9th session of the UNFCCC Conference of the Parties Meeting as the Parties to the Kyoto Protocol Opening Statement by the Like-Minded Developing Countries in Climate Change (LMDC)*	Nicaragua	11 November 2013
LMDC, 2013z. *LMDC Opening Statement for ADP – COP19*	Venezuela	12 November 2013
LMDC, 2013aa. *LMDC Statement in ADP on Technology Development and Transfer*	Egypt	14 November 2013
LMDC, 2013bb. *LMDC Statement on ADP Post-2020 Enhanced Action on Finance*	Egypt	14 November 2013
LMDC, 2013cc. *LMDC Statement on ADP Transparency of Action and Support*	Pakistan	15 November 2013
LMDC, 2013dd. *LMDC Statement on ADP Capacity-building*	Pakistan	15 November 2013
LMDC, 2013ee. *LMDC Statement on Workstream 2*	Venezuela	15 November 2013
LMDC, 2013ff. *LMDC Statement at ADP Informal Stocktaking Plenary*	Philippines	16 November 2013

(*Continued*)

Table 7.3 Continued

Source	Delivered by	Date
LMDC, 2013gg. *LMDC Views on Identification of Elements in ADP Workstream 1*	India	18 November 2013
LMDC, 2013hh. *LMDC Draft Decision Text Submitted to ADP Co-Chairs on Workstream 2*	N/A	18 November 2013
LMDC, 2013ii. *LMDC Statement for ADP Informal Plenary*	India	18 November 2013
LMDC, 2014a. *Submission on Elements of the 2015 Agreed Outcome*	Saudi Arabia, Ecuador, El Salvador, and Mali	8 March 2014
LMDC, 2014b. *LMDC Opening Statement – ADP 2.4*	Philippines	10 March 2014
LMDC, 2014c. *LMDC Closing Statements – ADP 2.4*	Venezuela	14 March 2014
LMDC, 2014d. *Proposal from the Like-Minded Developing Countries in Climate Change (LMDC). Decision X/CP.20 Elements for a Draft Negotiating Text of the 2015 ADP Agreed Outcome of the UNFCCC* [Conference Room Paper]	LMDC	3 June 2014
LMDC 2014e. *LMDC Opening Statement for ADP*	Egypt	4 June 2014
LMDC 2014f. *Proposal from the Like-Minded Developing Countries. Decision x/CP.20. Information on intended nationally determined contributions of Parties in the context of the 2015 agreed outcome* [Conference Room Paper]	LMDC	11 June 2014
LMDC 2014g. *Proposal from the Like-Minded Developing Countries in Climate Change (LMDC). Decision X/CP.20. Accelerating the implementation of enhanced pre-2020 climate action* [Conference Room Paper]	Jordan	21 October 2014
LMDC 2014h. *LMDC Statement on ADP Process*	Ecuador	25 October 2014
LMDC, 2014i. *LMDC Opening Statement – COP20*	Nicaragua	1 December 2014
LMDC 2014j. *LMDC Opening Statement – CMP10*	Pakistan	1 December 2014
LMDC, 2014k. *LMDC Opening Statement – ADP*	Cuba	2 December 2014
LMDC, 2015a. *Opening statement for ADP 2.8*	Bolivia	8 February 2015
LMDC, 2015b. *Opening statement for ADP 2.9*	Iran	1 June 2015
LMDC, 2015c. *Opening statement for ADP 2.10*	Malaysia	31 August 2015
LMDC, 2015d. *Statement concluding LMDC negotiators' meeting in India* [Press release]	India	15 September 2015

Table 7.4 List of interventions at High-Level Segment, COP19

Name	Title	Country	Date
Laila Rashed Iskander	Minister of Environment	Egypt	20 November 2013
Mary Ann Lucille Sering	Vice-Chairperson and Executive Director of the Climate Change Commission	Philippines	20 November 2013
Claudia Salerno	Vice Minister for Foreign Affairs	Venezuela	20 November 2013
Jayanthi Natarajan	Minister for Environment and Forests	India	21 November 2013
Vincent Makonga	Minister of Environment, Nature Conservation and Tourism	DRC	21 November 2013

Table 7.5 List of ENB sources

Source	Volume (number)	Pages
ENB, 2012a. Summary of the Bonn Climate Change Conference: 14–25 May.	12(546)	1–30
ENB, 2012b. Bangkok Climate Talks Highlights: Thursday, 30 August.	12(549)	1–4
ENB, 2012c. Doha Highlights: Monday, 3 December.	12(563)	1–4
ENB, 2012d. Summary of the Doha Climate Change Conference: 26 November–8 December.	12(567)	1–34
ENB 2013a, Summary of the Bonn Climate Change Conference: 29 April–3 May.	12(568)	1–19
ENB 2013b, Bonn Climate Change Conference: Monday, 3 June.	12(570)	1–2
ENB 2013c, Bonn Climate Change Conference: Tuesday, 4 June.	12(571)	1–4
ENB 2013d, Bonn Climate Change Conference: Wednesday, 12 June.	12(578)	1–2
ENB 2013e, Warsaw Highlights: Monday, 11 November.	12(584)	1–4
ENB 2013f, Warsaw Highlights: Monday, 18 November.	12(590)	1–2
ENB 2013g, Summary of The Warsaw Climate Change Conference: 11–23 November.	12(594)	1–32
ENB 2014, Summary of The Lima Climate Change Conference: 1–14 December.	12(619)	1–46
ENB 2015a, Summary of The Geneva Climate Change Conference: 8–13 February.	12(626)	1–16
ENB 2015b, Summary of The Bonn Climate Change Conference: 1–11 June.	12(638)	1–24
ENB 2015c, Summary of The Bonn Climate Change Conference: 31 August–4 September.	12(644)	1–16
ENB 2015d, Summary of The Bonn Climate Change Conference: 19–23 October	12(651)	1–11
ENB 2015e, Summary of The Paris Climate Change Conference: 29 November–13 December	12(663)	1–47

Notes

1 This chapter is largely adapted from a report commissioned prior to COP20 in 2014 (Blaxekjær et al., 2014).
2 The BASIC group, the Climate Vulnerable Forum (CVF), the Cartagena Dialogue for Progressive Action, the Durban Alliance (DA), the Mountainous Landlocked Developing Countries (MLDC), the Group of Like-Minded Developing Countries (LMDC), and Association of Independent Latin American and Caribbean States (AILAC).
3 The submitted statement from 25 May 2012 lists 38 countries: the Philippines on behalf of Algeria, Argentina, Bahrain, Bolivia, Comoros, China, Cuba, Democratic Republic of Congo, Dominica, Djibouti, Ecuador, Egypt, El Salvador, India, Iran, Iraq, Jordan, Kuwait, Libya, Malaysia, Mali, Mauritania, Morocco, Nicaragua, Oman, Palestine, Pakistan, Paraguay, Saudi Arabia, Somalia, Sri Lanka, Sudan, Syria, Thailand, Tunisia, Venezuela, and Yemen.
4 Algeria, Argentina, Bolivia, Cuba, China, Dominica, Ecuador, Egypt, El Salvador, India, Iran, Iraq, Kuwait, Nicaragua, Philippines, Qatar, Saudi Arabia, Sri Lanka, and Venezuela. In addition, Mali, Sudan, the DRC, Pakistan, and Thailand are included in at least four out of the explicit country lists on LMDC submissions.
5 Argentina, Bolivia, China, Cuba, El Salvador, Ecuador, Iran, Nicaragua, Venezuela, Malaysia, Vietnam, Saudi Arabia, and India.
6 Yeb Saño who is also famous for his emotional speech and fasting for two weeks at COP19 in Warsaw (Vidal, 2014).
7 Note: this statement was not delivered due to non-opening of SBI 38.

References

Bevir, M. (2005). How Narratives Explain. In D. Yanow & P. Schwartz-Shea (Eds.), *Interpretation and Method: Empirical Research Methods and the Interpretive Turn* (pp. 281–290). Armonk, NY: M.E. Sharpe, Inc.

Bidwai, P. (2014). The Emerging Economies and Climate Change: A Case Study of the BASIC Grouping. In *Shifting Power: Critical Perspectives on Emerging Economies.* Amsterdam: Transnational Institute.

Blaxekjær, L. Ø., Fang, F., Green-Weiskel, L., Kallbekken, S., Lahn, B., Nielsen, T. D., & Sælen, H. (2014). *Building Bridges with the Like-Minded Developing Countries.* Copenhagen: Nordic Working Group for Global Climate Negotiations (NOAK), Nordic Council of Ministers.

Blaxekjær, L. Ø., & Nielsen, T. D. (2015). Mapping the narrative positions of new political groups under the UNFCCC. *Climate Policy, 15*(6), 751–766.

Castro, P., & Klöck, C. (2020). Fragmentation in the Climate Change Negotiations: Taking Stock of the Evolving Coalition Dynamics. In C. Klöck, P. Castro, F. Weiler, & L. Ø. Blaxekjær (Eds.), *Coalitions in the Climate Change Negotiations.* Abingdon: Routledge.

Chan, N. (2020). The Temporal Emergence of Developing Country Coalitions. In C. Klöck, P. Castro, F. Weiler, & L. Ø. Blaxekjær (Eds.), *Coalitions in the Climate Change Negotiations.* Abingdon: Routledge.

Cobley, P. (2001). *Narrative.* Abingdon: Routledge.

Demeritt, D. (2001). The construction of global warming and the politics of science. *Annals of the Association of American Geographers, 91*(2), 307–337.

Fischer, F., & Forester, J. (1993). *The Argumentative Turn in Policy Analysis and Planning.* Durham, NC: Duke University Press.

Friedman, L. (2014, April 28). Can a diplomat who fasts on Earth Day deal with hard-liners at the bargaining table? *E&E News*. Retrieved from https://www.eenews.net/stories/1059998538.

Hochstetler, K., & Milkoreit, M. (2014). Emerging powers in the climate negotiations: Shifting identity conceptions. *Political Research Quarterly, 67*(1), 224–235.

Kasa, S., Gullberg, A. T., & Heggelund, G. (2008). The group of 77 in the international climate negotiations: Recent developments and future directions. *International Environmental Agreements: Politics, Law and Economics, 8*(2), 113–127.

Najam, A. (2005). Developing countries and global environmental governance: From contestation to participation to engagement. *International Environmental Agreements: Politics, Law and Economics, 5*, 303–321.

Nielsen, T. D. (2014). The role of discourses in governing forests to combat climate change. *International Environmental Agreements: Politics, Law and Economics, 14*, 265–280.

Pettenger, M. (Ed.). (2007). *The Social Construction of Climate Change: Power, Knowledge, Norms, Discourses*. Hampshire: Ashgate.

Pouliot, V. (2013). Methodology. In R. Adler-Nissen (Ed.), *Bourdieu in International Relations: Rethinking Key Concepts in IR* (pp. 45–58). London and New York: Routledge.

Rajamani, L. (2016). Ambition and differentiation in the 2015 Paris agreement: Interpretive possibilities and underlying politics. *International & Comparative Law Quarterly, 65*(2), 493–514.

Roberts, J. T. (2011). Multipolarity and the new world (dis)order: US hegemonic decline and the fragmentation of the global climate regime. *Global Environmental Change, 21*, 776–784.

RTCC. (2012). Interview with Nitin Desai, WWF. Retrieved from http://climatechange-tv.rtcc.org/2012/10/19/cbd-cop11-richpoor-divide-at-negotiations-is-here-to-stay-says-wwf/.

Sethi, N. (2013, November 7). India is not a nay-sayer on climate change. *The Hindu*. Retrieved from https://www.thehindu.com/sci-tech/energy-and-environment/india-is-not-a-naysayer-on-climate-change/article5323166.ece?homepage=true.

TWN. (2012, October 18–19). Meeting of the like-minded developing countries on climate change. Beijing, China [Press release]. Retrieved from http://www.twnside.org.sg/title2/climate/info.service/2012/climate20121005.htm.

TWN. (2013, February 27–28). Members of like-minded developing countries issue statement after an informal meeting in Geneva, Switzerland [Press release]. Retrieved from https://www.twn.my/title2/climate/info.service/2013/climate130301.htm.

TWN. (2015, December 4). Differentiation between developed and developing countries still relevant, say developing countries. *TWN Paris News Update*.

Vidal, J. (2014, April 1). Interview: Yeb Sano—Unlikely climate justice star. *The Guardian*. Retrieved from https://www.theguardian.com/environment/2014/apr/01/yeb-sano-typhoon-haiyan-un-climate-talks.

Wagenaar, H. (2011). *Meaning in Action: Interpretation and Dialogue in Policy Analysis*. Armonk and London: M.E. Sharpe.

8 One voice, one Africa

The African Group of Negotiators

Simon Chin-Yee, Tobias Dan Nielsen, and Lau Øfjord Blaxekjær

Introduction[1]

In the lead up to the Paris Agreement, there was considerable fear that one of the Parties to the United Nations Framework Convention on Climate Change (UNFCCC) would throw a spanner in the works and block the consensus needed for an agreement. The speculation was that an African country would wield this spanner. Indeed, in the hours before Laurent Fabius, President of the Conference of the Parties (COP) 21, declared the Paris Agreement adopted, he met with Ministers from Nigeria and Egypt to discuss unresolved issues. In the plenary itself, many countries had formed negotiating huddles or indabas, in particular around the South African delegation, deliberating the text right up to the moment the agreement was adopted. As we demonstrate in this chapter, Africa's status in the UNFCCC process has been on the rise over the past decade, even though their role in the negotiations has often been marginalised.

The African continent has a vested interest in the global climate regime and in particular since COP15 it has become a voice to be reckoned with in the climate negotiations. During that now infamous conference, Lumumba Stanislaus-Kaw Di-Aping, the then-Ambassador and Deputy Permanent Representative of Sudan to the United Nations in New York and the Chair of the G77 and China (G77), stated that "you cannot ask Africa to sign a suicide pact, an incineration pact in order to maintain the economic dominance of a few countries" (cited in Fisher, 2009).

This chapter examines African collective power and the influence African countries have in the global climate regime. This is achieved by analysing the African Group of Negotiators (AGN), this group's unique identity, its historical relevance, and how the AGN operates. We examine the AGN's unifying and dominant narrative of "One Voice, One Africa", while also highlighting key tensions, and discuss how the AGN cooperation has influence on national climate policy and African Nationally Determined Contributions (NDCs).

The chapter has two main sections. First, we present an in-depth analysis of the evolution and organisation of the AGN within the global climate regime, examining the rise of this group in recent years. To understand AGN as a coalition, we then analyse how the AGN has emerged as a voice in climate negotiations and has developed its identity and the dominant "One Voice, One Africa" narrative. We analyse the three central themes of AGN identity and narrative: vulnerability, responsibility, and representation. We conclude with a discussion section, emphasising the importance of the group as a regional negotiator, while recognising that you cannot apply this concept across the board, as each African country is culturally, economically, and socially distinct and needs to be examined individually.

Methodological approach

Our argument is underpinned through (1) elite interviews with policymakers and climate change specialists, (2) an analysis of policy developments and climate action plans and official statements, and (3) participant observation. A qualitative mixed methods framework was chosen due to the complexity of actor relationships within these networks of policymakers, and the variety of sources. This approach allowed the research to evolve and adapt within the negotiations of new rules and norms.

28 semi-structured elite interviews were conducted primarily at UNFCCC COPs and intersessional meetings (see the Appendix for full list). We interviewed policymakers, heads of country delegations, negotiators, and representatives of international agencies and civil society groups, as well as government advisors and climate policy academics/experts from a cross-section of African countries present in the negotiations. The official documents came in the form of statements from member states to the UNFCCC, both speaking on behalf of their country and/ or a negotiating group, statements from government officials, NDCs, UN reports, speeches, and civil society statements. Finally, participant observation during the UNFCCC negotiations was important in informing interactions between different actors. Participating in several of the COPs and intersessional meetings between 2014 and 2019 provided us with a broad view of state interactions with each other and the UNFCCC Secretariat (see full list in Appendix).

Throughout this chapter we apply a narrative analysis. Narrative, to us, is a *mode of knowing*, "providing distinctive ways of ordering experience, of constructing reality", such as an organisation's origin and identity (Wagenaar, 2011, p. 209).

The AGN

This section is an in-depth analysis of the evolution and rise of the AGN in the UNFCCC negotiations. We argue that the AGN has really come into its own since COP12 in Nairobi, and in particular in the lead up to the Paris Agreement in 2015. In order to understand the "One Voice, One Africa" narrative and identity, we start by describing the foundations of this negotiating coalition.

The evolution of the AGN

As a negotiating group, the AGN has a relatively short history. In its current form, the AGN only exists within the climate regime; however, it owes its existence to previous incarnations in the 1950s and 1960s, following decolonisation of the continent, promoting Pan-Africanism at the United Nations General Assembly (UNGA). In 1963 these groups merged to form the Africa Group of the Whole (AGW) to improve the coordination between states and consolidate continental-wide issues. Over time, similar Africa Groups started materialising in different UN agencies and forums, such as the negotiations on the Convention to Combat Desertification and the General Agreement on Tariffs and Trade. Perhaps most importantly, the AGW was responsible for selecting which African countries would be represented on the UN Security Council (Roger & Belliethathan, 2016, p. 94).

The AGN as we know it within the UNFCCC originated out of the Organisation of African Unity (OAU) Abuja Summit of Heads of State in 1991. It was recognised that in advance of the 1992 Rio Earth Summit the AGN needed to establish an "African common position on environment and development" to ensure that Africa's policies and concerns were adequately addressed (Najam, 2005, p. 133). This initial common position was the first time that the Africa Group addressed climate change, and it formed the basis of future common positions for the continent. Originally, the AGN, representing a continent minimally responsible for climate change but most vulnerable to its impacts, stressed the importance of economic development and poverty reduction over addressing environmental concerns, and upheld their sovereign right over the use of natural resources to tackle food and energy security (Roger & Belliethathan, 2016). Although their identity has remained the same, the AGN position has changed in very recent years as they now agree on the need to put in place national mitigation strategies, although their assertion of the need for financial aid and technology transfers to developing countries has never wavered.

It was during the 1992 UNFCCC meetings that the "Africa Group", or AGN, replaced the AGW. This new coalition reproduced the structure of the AGW; it was organised into five sub-regions and included all African negotiators. In the beginning, the AGN had trouble finding its feet amongst more established groups (such as the G77) and the complexities of the North–South negotiations.

An inherent strength (and weakness) of AGN representation in the negotiations is that although countries can have a strong voice, they gain more bargaining power when aligned with another group. In fact, many of the early AGN positions followed statements set out by the G77 and other negotiating groups that had a stronger leverage within the UN system. The AGN has had very little impact in the climate negotiations as compared to the G77 who helped shape the very structure of the global climate regime, including the role of Annex II countries in the 1995 Berlin Mandate (Roger & Belliethathan, 2016, p. 95). Even though AGN constitutes more than one-third of the countries that make up the G77, the AGN has historically relied heavily (whether they wanted to or not) on the larger emerging economies, such as China, Brazil, and India, who have a more powerful voice in both the G77 and the UNFCCC (Roger & Belliethathan, 2016, p. 96).

This marginal and largely powerless status has been slowly changing in recent years (Hurrell & Sengupta, 2012). The transformation of the AGN from a passive voice in the negotiations to a more vocal player began in earnest at COP12 Nairobi in 2006. This conference was to be known as the Africa Conference and was seen as an opportunity for the AGN to bring awareness to the plight of African countries. This was witnessed first-hand during the conference, as Kenya's Lake Nakuru and Lake Naivasha were completely dry and at a substantially lower level, respectively, the Horn of Africa was experiencing record droughts, and Nairobi experienced heavy flooding (Sterk, Ott, Watanabe, & Wittneben, 2007, p. 140).

The fact that COP12 and COP17 took place in Sub-Saharan Africa gave climate change more attention from African heads of state (own observations). Extra resources were also provided in order to host a preliminary meeting in September 2006. Through these extra efforts the African negotiators were able to come up with an African Common Position on adaptation, the Clean Development Mechanism (CDM), deforestation, climate finance, and technology transfer (Roger & Belliethathan, 2016, p. 99). During the conference, the AGN made it clear that for the continent, climate change and climate variability are inextricably linked to national development efforts. They called for enhanced human, institutional, and systematic capacity building initiatives, as well as adequate financial and technical assistance. They also challenged the distribution of CDM projects. At the time of COP12, only 15 CDM projects were under consideration in Sub-Saharan Africa, accounting for only 1.7% of all projects worldwide (Cosbey, Murphy, Drexhage, & Balint, 2006, p. 5).

In 2011, a lot was done to promote COP17 as an African COP, focus on vulnerability, and keep momentum from COP16 on e.g. finance (own observations). During COP17, African countries through the Alliance of Small Island States (AOSIS) and the LDCs, together with the European Union (EU), were instrumental in brokering the Durban Platform. This coalition was named the Durban Alliance (Blaxekjær & Nielsen, 2015). Fast-forward to COP21, Angola, The Gambia, and Ethiopia were among the countries who first presented the High Ambition Coalition, which with the support of more than 100 countries, many of them African LDCs, is credited to be the coalition that delivered the Paris Agreement (e.g. King, 2015; Mathiesen & Harvey, 2015; Observations at COP21).

Organisational structure of the AGN

Since COP12, the AGN has become more organised and better prepared and has been able to set the agenda on certain issues. For example, the AGN has daily planning meetings during the negotiations. It has also moved from acting as an autonomous entity to a more coordinated effort with other Pan-African organisations. AGN negotiations have direction from the African Union Assembly (AUA), the AU Committee of African Heads of State and Government on Climate Change (CAHOSCC), the African Ministers Conference of Environment Ministers (AMCEM), and the African Development Bank (AfDB), among others. Decisions of the AGN are officially made at the African Union (AU), under whose remit the AGN falls via AMCEM which, in turn, reports to the AU Assembly of Heads of State (Interview with Jarrett, 2015; Interview with Osman-Elasha, 2015).

The AGN is organised with a rotating chair (by region), it has a number of working groups with an experienced convenor and/or lead negotiator, and it receives technical and administrative support by international organisations such as ClimDev-Africa's The African Climate Policy Centre (ACPC), from NGOs and civil society (see Figure 8.1). All in all, the AGN has over time built up a strong and elaborate organisation. However, the AGN still faces several organisational challenges, including inadequate finances, limited capacity, lack of continuity of negotiators (institutional learning), and deficient strategic and technical assistance. It is also not clear to what degree there is tension between the AGN and the other (pan-African) organisations that all take part in formulating the AGN positions (see next section for more details).

The AGN's increasing visibility in the negotiations can be seen, at least in part, as having been facilitated by organisational improvements enabled by its own administration (see Figure 8.2). In addition, the skilled and experienced individual AGN negotiators receive both formal and informal input from different actors, including academics, scientists, and politicians, as well as assistance and backing from the AfDB and the Pan African Climate Justice Alliance (PACJA) (Interview with Ogallah, 2015; Interview with Osman-Elasha, 2015). Since 2011, the ACPC and the United Nations Economic Commission for Africa (UNECA) have organised an annual Climate Change and Development in Africa conference. According to the final report from the first conference, "Africa spoke with one voice (to the UNFCCC) and presented the continent's priorities as the global discussion on climate change mitigation and adaptation ensued" (UNECA, 2012, p. 6). Through using the two principal themes of vulnerability and responsibility, the AGN began to increase the third theme: representation. The next section will analyse the rise of the AGN and examine the strengths and weaknesses of this negotiating group.

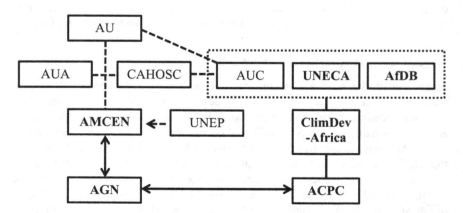

Figure 8.1 AGN external organisational structure (in relation to climate change). Adapted from Blaxekjær et al. (2015).

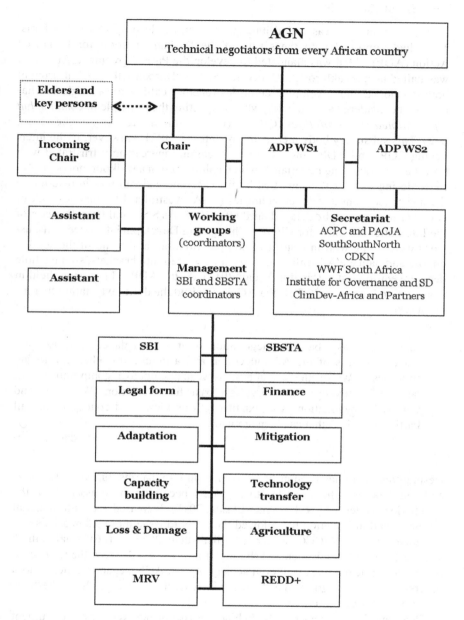

Figure 8.2 AGN internal organisation (in the lead-up to the Paris Agreement). Adapted from Blaxekjær et al. (2015).

The rise of the AGN

AGN representation was called into question in the lead-up to COP21 Paris. Talks had improved with the set-up of the Durban Platform for Enhanced Action (ADP) which was mandated to develop the Paris Agreement. At first, it was hailed as a breakthrough. Called for by the "Durban Alliance", it brought both developed and developing countries together calling for a single mechanism that considered all countries, while respecting the principle of *common but differentiated responsibilities* (CBDR) (Osafo, Sharma, & Abeysinghe, 2012, p. 3). This subsidiary body met regularly before COP21. However, in the intervening COPs and ADPs, the AGN was becoming increasingly frustrated with how the text was being negotiated, which culminated in a walkout, not once, but twice during COP19 Warsaw. Led by the G77, the walkout was in reaction to the insistence from developed countries (USA, Australia, EU) that any talk of compensation should be delayed until after 2015. As Saleemul Huq, Director of the International Centre for Climate Change and Development, stated, "discussions were going well in a spirit of co-operation, but at the end of the session on loss and damage Australia put everything agreed into brackets, so the whole debate went to waste" (cited in Vidal, 2013). At the ADP 2.11 session, held in Bonn a month prior to COP21, the AGN criticised the draft text, stating that the latest text

> cannot serve as a basis for negotiation, as it is not balanced and does not reflect concerns of the African Group and a number of other developing countries ... We see the text, at best, as weak attempt to a 'compromise text', rather than a text that could serve as a 'basis for negotiation'. We understand a basis for negotiation as a text that reflects views and concepts from all Parties in options that can be negotiated.
>
> (Sudan, 2015b)

Despite this, there has been an improvement in regional institutions, structures, and coordination, which has led to the AGN becoming more assertive in the negotiations (Interview with Osman-Elasha, 2015). Increased access to material resources and information has allowed the AGN to issue more submissions and statements to the UNFCCC (Roger & Belliethathan, 2016, p. 106). Finally, there is more cohesive organisation and sharing of information between the lead negotiators, scientific advisors, and civil society. As the AGN began to function as a progressively unified group, they began to increase their activity. This is the "One Voice, One Africa" narrative.

There are three deeper reasons behind the rise of the AGN: first, the urgent need for action at the international level and on the ground in Africa to effectively tackle the climate challenge. Second, the AGN has split in its negotiations from other groups. And, third, there are key individuals within the AGN who are young, impassioned, and determined to ensure that AGN positions are heard.

Urgency of tackling the climate challenge in Africa

African countries are particularly vulnerable to climate change and the need for capacity-building and technology transfer in Africa has often been highlighted during the negotiations. Ndonse (Interview, 2016), Burundi's national focal point for REDD (Reducing Emissions from Deforestation and Land Degradation), stated that the UNFCCC financial mechanism, such as "the GCF, the adaptation fund, the green fund can reinforce capacity that could translate to billions of dollars for national programmes". Silva (Interview, 2016), Angola's Head of Vulnerability Department, Direction of Climate Change in the Ministry of Environment, reinforced this view by stressing that Angola's 2011 National Adaptation Plan of Action (NAPA) was never implemented because of finance, but that through the Convention and joint ventures with UNEP and FAO, they are receiving training that includes [1] help to develop guidelines for the National Adaptation Plan (NAP) process as well as [2] producing guides to help access the funds. A common theme is the need for in-country capacity-building, technology transfer, and financial assistance for policy development and implementation. In order to put their positions on these matters forward, many countries look to the AGN to be their voice.

Split in priorities between the AGN and other key coalitions

The moment that the AGN began to have a more prominent voice was when their relationship to other groups changed, as their positions began to differ, most notably with the G77 (Interview with Jarrett 2015). Additionally, given the proliferation of coalitions since 2009 (see Castro & Klöck, 2020), AGN countries have yet more coalitions to align to, further complicating the negotiating process in the global climate regime (see Figure 8.3).

For example, the AGN and the LDCs were at odds with the larger developing countries on CDM, a market-based instrument allowing developed countries to offset their emissions by subsidising sustainable development in developing countries, often overlooking African countries. This division was further reinforced with the establishment of the BASIC group (a bloc of the four large industrialised countries who are also member of the G77: Brazil, South Africa, India, and China). It became clear that it is in the interest of the AGN (excluding South Africa) for these large emitters to have mitigation targets, as they are increasingly responsible for a large share of global CO_2 emissions. These four countries often present themselves as the voice of the developing world, as they wield enormous power within the G77; but they can end up competing with much poorer countries for CDM adaptation and mitigation projects (van Schaik, 2012; Interview with Souleymane, 2015; Interview with Elie, 2015). As such, for many African countries it is more fitting to align themselves with the AGN or LDCs.

In the negotiations prior to COP21, the AGN has made more submissions (15) than any other group.[2] They made six submissions from May to October 2015 alone indicating that the final text of what was to become the Paris Agreement was

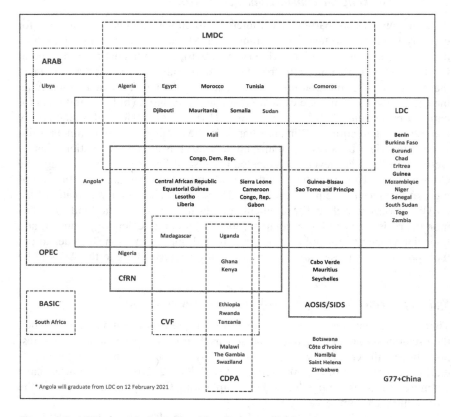

Figure 8.3 AGN countries' membership of other coalitions.

central for the AGN. Individual African states, in contrast, have been rather inactive; only South Africa (5), Algeria (2), and Ethiopia (1) have made individual submissions in the lead up to COP21. This is an indication that Africa is speaking with a united voice, and evidence that most countries on the continent let the AGN speak on their behalf.

Personal drivers behind the AGN: negotiators, policymakers, and heads of state

In the last decade, key individuals within the AGN have become very proactive. Compared to other coalitions the number of negotiators is very small, but they are determined that the AGN will not take a back seat during the negotiations (Interview with Osman-Elasha, 2015). They range from policy advisors, the negotiators themselves, the Chair of the AGN, the Ministries, and the heads of state. They ensure the internal functioning of the group as well as protecting their own national interests. Emmanuel Dlamini from Swaziland, the AGN Chair from 2012–2013 reflected that

> Africa is beginning to have a strong voice in the negotiations that needs to be maintained and strengthened ... The AGN undertook a rigorous process in preparing for the 19th Conference of Parties (COP) ... Preparations were challenging due to regional diversity in development, language, culture, geography, and relative vulnerability to climate change. Despite these constraints, as the AGN Chair I managed to pull together over 50 members of the AGN for a preparatory meeting in the Kingdom of Swaziland.
>
> (Dlamini, 2014)

Certain key figures hold a prominent place in the functioning and positions of the AGN. Yamin and Depledge (2004) note that many African country representatives do not have the experience to take on the role of lead coordinator. One of the reasons is the high turnover of delegates from certain countries, which undermines consistency and institutional learning (Interview with Damptey, 2015). This makes the few central figures that have participated regularly in the negotiations all the more significant to the AGN. Moreover, the very active diplomatic engagement of past leaders like Meles Zenawi and Jacob Zuma, and the willingness of Ethiopia and South Africa to play high-profile leadership roles in climate diplomacy have increased African influence and power (Brown & Harman, 2013).

Balgis Osman-Elasha (Interview, 2015), Climate Change Expert, Compliance and Safeguards Division of the Africa Development Bank, explained that the AGN is supported by several organisations, including the Bank, and together they identified "gaps in knowledge" and developed technical support for the group, in particular supporting the lead coordinators. These joint efforts have proved very fruitful and the AGN went to Paris with solid positions.

Building the AGN identity

This section contextualises the African perspective on climate negotiations and unpacks the dominant narrative of the AGN with a focus on three themes of AGN identity: vulnerability, responsibility, and representation.

Vulnerability

Vulnerability is a key element in the AGN narrative and thus its identity. For Africa, climate change is as much about climate justice as it is about a changing environment. For the AGN, climate change disasters are no longer simply isolated weather events and/or related to a lack of knowledge and government preparedness (Varley, 1994, p. 3), but an accurate depiction of inequalities that exist between developed and developing countries.

It is widely recognised that Africa will be hit faster and more strongly by a changing climate than many other parts of the world (Adger, Huq, Brown, Conway, & Hulme, 2003, p. 184; Collier, Conway, & Venables, 2008, p. 337; Pelling & Uitto, 2001, p. 55). The effects are already being seen. Throughout the continent, climate-related impacts, such as droughts, failing crops, unpredictable weather patterns, and rising temperatures in lakes have escalated in recent years (Collier et al., 2008, p. 339). The resulting economic instability and food insecurity have led to large-scale migration and displacement of communities and conflict over the remaining resources (Black et al., 2011, p. S7; Brown, Hammill, & McLeman, 2007, p. 1149; Chin-Yee, 2019, p. 15). This "new" unstable environment gives rise to security threats at the national and transnational level. The 2011 Somali famine shows a clear link between drought and conflict, which has resulted in a significant increase in Somali refugees in Djibouti, Ethiopia, and Kenya, with an estimated 1.5 million people internally displaced (Ferris, 2012, p. 4; Ferris & Petz, 2011, p. 104). Desertification is having direct socio-economic and cultural impact on populations (Interviews with: El Wavi, 2018; Manyok, 2016 & 2018; Source 7, 2018; Source 8, 2018).

Since the creation of the ADP (the UNFCCC body in charge of negotiating the Paris Agreement) in 2011, the AGN has been lobbying for an agreement that would hold global temperature rises to 1.5°C, rather than the divisive 2°C that was agreed on in Copenhagen (Interview with Jarret, 2015), creating a higher sense of vulnerability in Africa and a sense of urgency that needs to be met with concrete action on the ground.

Responsibility

The second theme is responsibility, of which the African continent has little when it comes to anthropogenic climate change. Africa accounts for less than 4% of total global greenhouse gas emissions (Sy, 2015, p. 58). Whether looking at per capita emissions or production and consumption-related emissions, it has been argued fervently that the African Continent is not historically responsible (Interview with Ogallah, 2015; Observations ADP 2.9, COP21). The AGN position has always placed great emphasis on the principles of CBDR and climate justice. However, carbon emissions are not uniform across the African continent. There are petroleum-reliant economies such as Angola, Sudan, and Nigeria. Even Africa's two largest emitters and economies, Nigeria and South Africa, have significant disparities between them. Despite having one-third of the population, South Africa uses nine times more energy than Nigeria (Africa Progress Panel, 2015). Disparities exist also at the level of policies and legal frameworks, particularly for managing oil and gas resources (Africa Development Bank & African Union, 2009, p. 182).

Canadell, Raupach, and Houghton (2009) analyse the difference between carbon emissions through fossil fuels and land use. It concluded that in Africa, 48% of emissions came from deforestation and land use primarily by Central African countries, such as the Democratic Republic of the Congo, Nigeria, and Zambia, whereas emissions from fossil fuels were dominated by South Africa and the North African countries. The study also noted that even with rapid population growth and an increase in GDP per capita, African CO_2 emissions will still remain comparatively minor as compared to other continents (Canadell et al., 2009). In fact, Africa has considerably lower average per capita energy consumption than other world regions (World Wildlife Fund for Nature & Africa Development Bank, 2012, p. 37).

One crucial step that the AGN is promoting, in line with CBDR, is the importance of addressing the needs of the ever-growing populations in combination with tackling climate change at the national level. Through much of the AGN narrative the lines between sustainable development and climate change have become blurred. In a statement to the ADP 2.9 by Sudan on behalf of the AGN, they made it clear that they support renewable energy projects in Africa, linking the "real economy world" with the negotiations. They stated

> Today, the 54 countries of Africa have around 1 billion inhabitants and the continent's population will grow to around 2 billion by 2050. Ensuring the wellbeing of these people requires, among other things, access to reliable and affordable clean energy. Yet, today as many as 600 million people lack adequate access to such energy. To achieve its full potential, Africa must rapidly scale up access to energy, to meet the needs of an expanding population, while also curbing the growth of greenhouse gas emissions to address the threat of climate change which, if left unchecked, will undermine development.
>
> (Sudan, 2015a)

Responsibility is then framed as a need for Africa to address the energy needs of its ever-expanding population. However, according to the Africa Progress Panel (2015, p. 68), "Africa's energy systems stand at a crossroads". Despite the abundance of renewable energy sources, Africa have yet to realise this potential (Africa Development Bank & African Union, 2009, p. 198), as fully "[t]wo-thirds of the energy infrastructure that should be in place by 2030 has yet to be built" (Africa Progress Panel, 2015, p. 68). Africa may not be historically responsible for the anthropogenic climate change, but it can help drive change. With the right support, African countries have the opportunity to put in place policies that addresses climate change, provide much needed energy security, and grow its economic capacities.

Representation

The third theme is representation. As one of the most vulnerable and least historically responsible regions for anthropogenic climate change, Africa carries a certain moral legitimacy into the negotiations. However, what weight does this actually carry? Since the inception of the UNFCCC in 1992, African countries have traditionally had little bargaining power in the negotiations. They can neither offer major emission reductions nor financial support. In need of financial assistance to implement their climate activities, their position has traditionally been cast as "passive recipients of climate diplomacy rather than active participants" (Chin-Yee, 2016, p. 360). This has improved with the AGN becoming more vocal in the past decade, despite some countries not being able to send negotiators to the important intersessionals.

The negotiating strategy of the AGN is to maximise both their moral legitimacy and their strength in numbers. As one of the largest and most coherent blocs within the UNFCCC process, the attempt to maintain a common African position gives the AGN a certain weight and prominence (Interview with Mpanu Mpanu, 2015). Its purpose is to promote its members' collective interests, enhancing their capacity within the climate negotiations in the UNFCCC. As the AGN has evolved, it is changing how African countries are represented in the global climate regime.

Conclusion and discussion

As illustrated in this chapter, the one voice narrative of the AGN is based on three recurring themes, *responsibility, vulnerability, and representation.* Cultural, social, and economic diversity notwithstanding, the AGN is not just held together by geography, but also a shared post-colonial history and narrative, making it a less fractured group than other groups, such as the G77. This context matters. The ability for an individual African country to have real influence in the climate negotiations has been questioned, no less than by African countries themselves. As a result, there is a feeling that Africa needs to stick together.

Moreover, the importance of the AGN is highlighted by the influence it has on national policymaking in Africa. For example, the AGN influenced Zimbabwe's climate change response strategy, mirroring the AGN stance on the Paris Agreement (Interview with Source 2, 2016).

However, counterbalancing this dominant "One Voice, One Africa" narrative is an alternative narrative that the AGN does not represent (all of) Africa. In concluding this chapter, it would be remiss of us to not highlight this alternative or parallel narrative. A narrative that suggests that the AGN is more a tacit union, as opposed to a collective agreement, where individual African countries may not agree or abide with AGN positions.

Covering 54 countries, all of whom belong to other, sometimes multiple, coalitions, the AGN represents a very diverse set of member states in terms of political, economic, social, and cultural characteristics (Interview with Mpanu Mpanu, 2015). Although the AGN negotiates with one voice, not all African countries agree to the dominant narrative and the solutions put forward, pointing to the existence of an alternative narrative. The explanation of this alternative narrative is found chiefly in two situations: first, that the AGN, despite its many united actions, statements, and submissions, does not equally represent all 54 countries, and that countries, while letting the AGN represent them in the negotiations, do not in all instances agree with AGN positions. Second, the current incarnation of the AGN has had to grow up quickly and concurrently with many other changing coalition dynamics in the UNFCCC, which has provided many countries with a plethora of options in relation to joining new coalitions. For certain countries with fewer or limited resources it may seem more beneficial to keep the status quo, and lean on the usual and more resourceful partners, such as different negotiating coalitions (LDCs, AOSIS, Arab Group, etc.) and/or keep relations with traditional donor countries and organisations.

As resources differ from country to country, individual countries build relationships to coalitions that are more aligned to their issues (Interview with Jumeau, 2018). Overlapping coalition memberships can cause tensions (Castro & Klöck, 2020) and presents an alternative narrative to the dominant AGN "One Voice, One Africa" narrative. AGN countries make up more than one-third of countries in the G77, which presents a very clear challenge from the outset: the AGN is both a single "entity" with its own membership and positions, as well as a collection of states that are also members of other coalitions with other, often overlapping, but sometimes different positions (Interview with Roger, 2015). Because of overlaps in membership with other coalitions, core positions within the AGN tend to fluctuate. Countries will align themselves with a particular platform that best represents them (Interview with Roger, 2015).

The AGN can be seen as having a set of "ready-made" institutional resources and alliances that countries access when it is more tactically useful for them to go to the negotiating table as part of the AGN, and conversely when it would be more advantageous for them to align themselves with another group. This is problematic when attempting to measure the AGN's power and negotiating tactics, as it has to be assessed in relation to other groups and specific issues. As a result of this, there is a constant balancing act between interests and resources of individual coalitions and countries (Interview with Jumeau, 2018). Understanding how individual African states are reacting to and working with the global climate change regime is fundamental to understanding AGN negotiation tactics.

That being said, this chapter has focused on the collective response of the AGN within the UNFCCC processes. We argue that it is this negotiating coalition that speaks on behalf of the African continent on alienating changes in the negotiated text (for example, the virtual removal of *loss and damage* in the Paris Agreement was quite a blow for the AGN as it symbolises responsibility), that reacts to positions by other parties and groups during the plenaries (through showing support for action taken by another group), and that promotes the African position with regards to action within the UNFCCC process (pushing for financial assistance to developing countries). It seeks to build cohesion and strength within the continent and arrive as "one voice".

So, what can we learn from the "One Voice, One Africa" story? The AGN has undoubtedly been a positive force in moving climate change policy processes in Africa forward. As a centralising voice, the AGN has the ability to influence the global climate regime, which necessarily impacts national climate change policy for Africa. As a particularly vulnerable continent that could potentially have all its recent development gains wiped out in the decades to come as a result of rising temperatures, African negotiators realised that its climate governance is a matter of life and death. Even though the AGN has recently gained some power within the global climate regime, the fact of the matter is that African negotiators have had to battle to ensure that African priorities are reflected in the international negotiations (Hurrell & Sengupta, 2012; Roger & Belliethathan, 2016). However, it is no longer true to assert that "the African continent faces difficulties in getting its voice heard in global arenas" (Toulmin, 2009, p. 149).

The AGN is vital in assisting countries in negotiating funds and mechanisms within the global climate regime that will allow them to access critical financial support for policy implementation. Capacity-building is also very important in developing climate action plans and policy. For many African governments, the inclusion of a department focused on climate change is relatively recent, which results in greater attention to AGN positions. The AGN takes on a whole new importance for countries that are under-represented. This is particularly true for countries, for example, Eritrea or the Central African Republic, whose governments need to focus their energy on other priorities. The AGN provides substance for national policy as well as an effective negotiating voice in the global climate regime.

The AGN needs to assess the Paris Agreement by how far-reaching and ambitious it is globally, coupled with African interests being represented. It is a result of direct pressure from, amongst others, the AGN that Article 2 of the Paris Agreement holds countries "to pursue efforts to limit the temperature increase to 1.5°C above pre-industrial levels" (UNFCCC, 2015). However, Africa has lost its special status, the Paris text singling out only the LDCs and SIDS for particular support and flexibility, which means that countries such as Kenya no longer fall into this category (Chin-Yee, 2016, p. 367). On the other hand, taking a bottom-up approach, the agreement does not impose particular models of development or growth paths on the continent, which falls in line with the AGN Common Position, recognising the notion of CBDR.

As we have stated repeatedly, the AGN approaches the global climate regime with the moral legitimacy afforded to it by the three core themes of responsibility, vulnerability, and representation. This should not be underestimated; it forms the central character of the "One Voice, One Africa" narrative, and addresses the inequalities that have plagued African countries since colonisation. At the end of COP21, African country representatives could hold their heads up high and go back to their countries holding an agreement that everyone agreed is not the "perfect" document but is a place to start building their national capacities and transforming their NDCs into action.

Appendix

Table 8.1 to 8.3

Table 8.1 Events observed by one or more of the co-authors

Name of event	Place	Time
COP20/CMP10	Lima, Peru	1–13 December 2014
ADP2-8	Geneva, Switzerland	8–13 February 2015
ADP2-9	Bonn, Germany	1–11 June 2015
COP21/CMP11	Paris, France	30 November–12 December 2015
COP22/CMP12	Marrakech, Morocco	7–18 November 2016
COP23/CMP13	Bonn, Germany	6–17 November 2017
SB48	Bonn, Germany	30 April–10 May 2018
COP24/CMP14	Katowice, Poland	3–14 December 2018
COP25/CMP15	Madrid, Spain	2–13 December 2019

Table 8.2 Interviews conducted pre-COP21

No.	Interviewees and affiliation	Interview date
1	**Jackson Kiplagat**, Governance Coordinator, WWF: Kenya Country Office (KCO)	9 February 2015
2	**Source 3**, Horn of Africa Delegation	9 February 2015
3	**Patience Damptey**, Consultant, Gender and Environment for Ghana	10 February 2015
4	**Nsiala Tosi Bibanda Mpanu Mpanu**, Former Chair of the AGN, Conseiller Principal en charge de l'Environnement près 1er Ministre for the Democratic Republic of the Congo	11 February 2015
5	**Max Bankole Jarrett,** Deputy Director, Africa Progress Panel	25 March 2015
6	**Charles Roger**, University of British Colombia	7 May 2015
7	**Source 1**, Central African Delegation	4 June 2015
8	**Source 10**, Pan Africa Climate Justice Alliance	4 June 2015
9	**Chebet Maikut**, Commissioner at the Climate Change Department, Ministry of Water and Environment and UNFCCC National Focal Point for Uganda	4 June 2015
10	**Source 4**, Climate Change Expert and Legal Advisor, African Group of Negotiators (AGN)	8 June 2015
11	**Source 5**, Senior Government Official	8 June 2015
12	**Source 6**, Government Official	8 June 2015

Table 8.3 Interviews during COP21 and after

No.	Interviewees and affiliation	Interview date
13	**Samson Samuel Ogallah,** Programme Manager, Pan African Climate Justice Alliance	9 December 2015
14	**Hamid Abakar Souleymane,** Chad's Director of Exploitation and of the Meteorological Applications, National Focal Point with the IPCC and 2nd National Focal Point with UNFCCC	10 December 2015
15	**Mbaitoubam Elie,** Director General of the National Meteorology, PR of Chad with WMO	10 December 2015
16	**Balgis Osman-Elasha,** Climate Change Expert, Compliance and Safeguards Division of the Africa Development Bank	10 December 2015
17	**Source 2,** Southern African Delegation	9 November 2016
18	**Payai John Manyok,** National Focal Point for UNFCCC/ MEAs Officer at Ministry of Environment and Forest, Juba South Sudan	9 November 2016
19	**Sylvestre Ndose,** National Focal Point for REDD+ for Burundi	14 November 2016
20	**Cecilia Silva,** Angola's Head of Vulnerability Department, Direction of Climate Change in the Ministry of Environment	14 November 2016
21	**Seth Osafo,** Consultant, International Environmental Law, former senior legal adviser of the UNFCCC Secretariat	15 November 2016
22	**Ronnie Jumeau,** Ambassador, the Permanent Mission of the Republic of Seychelles to the United Nations	3 May 2018
23	**Source 7,** West African Delegation	4 May 2018
24	**Source 8,** West African Delegation	4 May 2018
25	**Sidi Mohamed El Wavi,** Chargé de Mission | Cabinet du Ministre de l'Environnement et du Développement Durable, Coordonnateur du Programme National sur le Changement Climatique (CCPNCC) – République Islamique de Mauritanie	7 May 2018
26	**Source 9,** West African Delegation	7 May 2018
27	**Petrus Muteyauli,** Deputy Director, Ministry of Environment and Tourism, Department of Environmental Affairs, Namibia	7 May 2018
28	**Michel Omer Laivao,** Point Focal National Changement Climatique, Madagascar	8 May 2018

Notes

1 This chapter largely builds on the chapter "One Voice, One Africa" in Chin-Yee (2018) and a report commissioned prior to COP21 in 2015 (Blaxekjær, et al., 2015).
2 Number of ADP submissions by selected groups (January 2012 to October 2015): AGN (15), EU (14), LMDC (10), EIG (9), AILAC (8), LDC (6), AOSIS (5), and UG (2) (Blaxekjær et al., 2015).

References

Adger, W. N., Huq, S., Brown, K., Conway, D., & Hulme, M. (2003). Adaptation to climate change in the developing world. *Progress in Development Studies, 3*(3), 179–195.

Africa Development Bank, & African Union. (2009). *Oil and Gas in Africa*. Oxford: Oxford University Press.

Africa Progress Panel. (2015). *Power, People, Planet: Seizing Africa's Energy and Climate Opportunities*. Abeokuta: Africa Progress Group.

Black, R., Adger, W. N., Arnell, N. W., Dercon, S., Geddes, A., & Thomas, D. S. G. (2011). The effect of environmental change on human migration. *Global Environmental Change, 21*, S3–S11.

Blaxekjær, L. Ø., Chin-Yee, S., Kallbekken, S., Nielsen, T. D., & Sælen, H. (2015). *Building Bridges with the African Group of Negotiators*. Copenhagen: Nordic Working Group for Global Climate Negotiations (NOAK), Nordic Council of Ministers.

Blaxekjær, L. Ø., & Nielsen, T. D. (2015). Mapping the narrative positions of new political groups under the UNFCCC. *Climate Policy, 15*(6), 751–766.

Brown, O., Hammill, A., & McLeman, R. (2007). Climate change as the "new" security threat: Implications for Africa. *International Affairs, 83*(6), 1141–1154.

Brown, W., & Harman, S. (Eds.). (2013). *African Agency in International Politics*. Abingdon: Routledge.

Canadell, J. G., Raupach, M. R., & Houghton, R. A. (2009). Anthropogenic CO_2 emissions in Africa. *Biogeosciences, 5*(6), 463–468.

Castro, P., & Klöck, C. (2020). Fragmentation in the Climate Change Negotiations: Taking Stock of the Evolving Coalition Dynamics. In C. Klöck, P. Castro, F. Weiler, & L. Ø. Blaxekjær (Eds.), *Coalitions in the Climate Change Negotiations*. Abingdon: Routledge.

Chin-Yee, S. (2016). Briefing: Africa and the Paris climate change agreement. *African Affairs, 115*(459), 359–368.

Chin-Yee, S. (2018). *Defining climate policy in Africa: Kenya's climate change policy processes* (PhD thesis). University of Manchester, Manchester.

Chin-Yee, S. (2019). *Climate Change and Security: Linking Vulnerable Populations to Increased Security Risks in the Face of the Global Climate Challenge*. European Centre for Energy and Resource Security and Konrad Adenauer Foundation.

Collier, P., Conway, G., & Venables, T. (2008). Climate change and Africa. *Oxford Review of Economic Policy, 24*(2), 337–353.

Cosbey, A., Murphy, D., Drexhage, J., & Balint, J. (2006). *Making Development Work in the CDM: Phase II of the Development Dividend Project*. Winnipeg: International Institute for Sustainable Development.

Dlamini, E. (2014). Opinion: Former AGN chair reflects on representing a strong African voice in climate negotiations. Retrieved from https://cdkn.org/2014/01/opinion-former-agn-chair-reflects-on-representing-a-strong-african-voice-in-climate-negotiations/?lo clang=en_gbv

Ferris, E. (2012). *Internal Displacement in Africa: An Overview of Trends and Opportunities*. Paper presented at the Ethiopian Community Development Council annual conference "African Refugee and Immigrant Lives: Conflict, Consequences, and Contributions". Retrieved from https://www.brookings.edu/wp-content/uploads/2016/06/0503_displac ement_africa_ferris.pdf

Ferris, E., & Petz, D. (2011). *The Year that Shook the Rich: A Review of Natural Disasters in 2011*. Washington, DC: Brookings Institution.

Fisher, A. (2009, December 19). Little accord in Copenhagen. *Al Jazeera*. Retrieved from https://www.aljazeera.com/focus/climatesos/2009/12/20091219174523761297.html

Hurrell, A., & Sengupta, S. (2012). Emerging powers, North—South relations and global climate politics. *International Affairs, 88*(3), 463–484.

King, E. (2015, December 14). Foie gras, oysters and a climate deal: How the Paris pact was won. *Climate Change News*. Retrieved from https://www.climatechangenews.com /2015/12/14/foie-gras-oysters-and-a-climate-deal-how-the-paris-pact-was-won/

Mathiesen, K., & Harvey, F. (2015, December 8). Climate coalition breaks cover in Paris to push for binding and ambitious deal. *The Guardian*. Retrieved from https://www.the guardian.com/environment/2015/dec/08/coalition-paris-push-for-binding-ambitious-climate-change-deal

Najam, A. (2005). Developing countries and global environmental governance: From contestation to participation to engagement. *International Environmental Agreements: Politics, Law and Economics, 5*, 303–321.

Osafo, S., Sharma, A., & Abeysinghe, A. C. (2012). *Durban Platform for Enhanced Action: An African Perspective*. European Capacity Building Initiative. Oxford.

Pelling, M., & Uitto, J. I. (2001). Small island developing states: Natural disaster vulnerability and global change. *Environmental Hazards, 3*, 49–62.

Roger, C., & Belliethathan, S. (2016). Africa in the global climate change negotiations. *International Environmental Agreements: Politics, Law and Economics, 16*(1), 91–108.

Sterk, W., Ott, H. E., Watanabe, R., & Wittneben, B. (2007). The Nairobi climate change summit (COP 12 – MOP 2): Taking a deep breath before negotiating post-2012 targets? *Journal for European Environmental and Planning Law, 4*(2), 139–148.

Sudan. (2015a). *Statement on Behalf of the African Group of Negotiators (AGN) by the Republic of the Sudan, at the Opening Plenary of Ninth Part of the Second Session of the Ad Hoc Working Group on Durban Platform for Enhanced Action (ADP 2.9)*. Bonn.

Sudan. (2015b). *Statement on Behalf of the African Group of Negotiators, by the Republic of Sudan, at the Opening Plenary of the Eleventh Part of the Second Session of the Ad Hoc Working Group on the Durban Platform for Enhanced Action*.

Sy, A. (2015). *Africa: Financing Adaptation and Mitigation in the World's Most Vulnerable Region*. Washington, DC: Brookings Institution.

Toulmin, C. (2009). *Climate Change in Africa*. London: Zed Books.

UNECA. (2012). The First Conference on Climate Change and Development in Africa. In *ACPC Technical Report*. Addis Ababa: United Nations Economic Commission for Africa.

UNFCCC. (2015). Paris Agreement. In *Contained in Document FCCC/CP/2015/10/Add.1*.

van Schaik, L. (2012). *The EU and the Progressive Alliance Negotiating in Durban: Saving the Climate?* London: Overseas Development Institute and Climate and Development Knowledge Network.

Varley, A. (1994). The Exceptional and the Everyday: Vulnerability in the International Decade for Disaster Reduction. In A. Varley (Ed.), *Disaster, Development and Environment* (pp. 1–12). Chichester: Wiley.

Vidal, J. (2013, November 20). Poor countries walk out of UN climate talks as compensation row rumbles on. *The Guardian*. Retrieved from https://www.theguardian.com/global-development/2013/nov/20/climate-talks-walk-out-compensation-un-warsaw

Wagenaar, H. (2011). *Meaning in Action: Interpretation and Dialogue in Policy Analysis*. Armonk and London: M.E. Sharpe.

World Wildlife Fund for Nature, & Africa Development Bank. (2012). *Africa Ecological Footprint Report: Green Infrastructure for Africa's Ecological Security*. Gland and Tunis Belvedere: WWF – World Wide Fund for Nature and AfDB – African Development Bank.

Yamin, F., & Depledge, J. (2004). *The International Climate Change Regime: A Guide to Rules, Institutions and Procedures*. Cambridge: Cambridge University Press.

9 AILAC and ALBA

Differing visions of Latin America in climate change negotiations

Joshua Watts

Introduction: Latin America in the climate change negotiations

This chapter provides an introduction to the two explicitly Latin American nego-tiating coalitions within the United Nations Framework Convention on Climate Change (UNFCCC), the Bolivarian Alliance for the Peoples of our America (ALBA)[1] and the Independent Association of Latin America and the Caribbean (AILAC). If Latin America can be understood as a "bellwether for the future of international climate change negotiations because its countries are a diverse, con-structive, and controversial group of actors at the UN climate talks" (Edwards & Roberts, 2015a, p. 35), AILAC and ALBA appear to occupy opposite ends of this diverse Latin American spectrum in the negotiations. They demonstrate sharply differing negotiating positions and manifest different behaviours, despite sharing many of the characteristics identified in Chapter 1 as defining climate negotiating groups.

There has been relatively sparse analysis of ALBA and AILAC since their emergence in 2009 and 2012 respectively,[2] part of a larger research gap: despite the rich influence of Latin American states within the climate regime "[s]ince the birth of the UNFCCC in 1992, little academic literature has looked at Latin America and climate change" (Edwards & Roberts, 2015a, p. 35). In this chapter, insight is therefore drawn from the wider literature on climate negotiating groups, and from analysis of other regional cooperation projects in Latin America. As an overview, this chapter draws on publicly available primary sources, namely, reports from the Earth Negotiations Bulletin (ENB) and COP/subsidiary body plenary statements. Throughout the chapter, emphasis is given to the vital con-text attained by setting ALBA and AILAC's activity and positions in the climate negotiations against the backdrop of Latin America's long and contested tradition of regional projects, echoing Puntigliano's (2013, p. 21) assertion that in South America "to understand current regional expressions and analyze their strategic scope one has to grasp the waves of regionalisms that preceded them".

In this contribution to the relatively scarce literature on Latin American cli-mate negotiating coalitions the two groups are examined and compared along the three dimensions identified by Castro and Klöck (2020) in the introduction to this volume: scope, membership, and formality. Considering the group's similarities and differences along these dimensions casts into relief the greater level of engagement and cohesion which AILAC appears to demonstrate within the climate negotiations.

Scope and membership

As defined by Castro and Klöck (2020), a coalition's scope can be understood in two senses: as a geographic pre-qualification for membership, and in a thematic sense, capturing the range of issues on which the group is active. In both senses there is considerable overlap, and therefore contestation, between the AILAC and ALBA coalitions.

Geographic scope

Both groups are exclusively Latin American, placing them firmly in the category of a negotiating coalition with a clear geographical scope. AILAC and ALBA are the two groups within the negotiations which seek to explicitly represent Latin America. This configuration was absent for the first two decades of the climate negotiations, prior to ALBA's emergence; two sub-regional groupings, the Central American Integration System (SICA: *Sistema de Integración Centroamericana*) and the Caribbean Community (CARICOM), have been longstanding actors in the climate negotiations, but have tended to play a mostly operational role focused on climate change responses on the ground, rather than on political negotiation (although this may be changing). Likewise, as Castro and Klöck (2020) note, in the climate change negotiations, as in most issue-specific processes, Latin America's regional group in the UN system (Group of Latin America and Caribbean Countries, GRULAC) is used almost exclusively for elections to regime bodies (notably the COP Bureau) and similar purposes, not for substantive negotiation. The creation of AILAC and ALBA, as negotiating coalitions based explicitly on regional identity and Latin American geographical scope, therefore constituted a new and significant departure for Latin American representation in the climate change negotiations.

Whilst AILAC and ALBA are explicitly and exclusively Latin American negotiating groups in the UNFCCC, other countries in the region, and indeed countries within the two groups, are active in an extremely wide range of negotiating groups, represented below in Figure 9.1. This heterogeneity prompted Edwards and Roberts's (2015a, p. 37) observation that "[l]umping together Latin America's diverse countries into one monolithic region is unhelpful. In over twenty years of international climate negotiations, Latin American countries have rarely spoken with one voice". Significantly Latin America's two largest economies, Brazil and Mexico, have looked beyond the region for negotiating allies. Mexico forms part of the Environmental Integrity group (EIG) along with Liechtenstein, Georgia, Monaco, South Korea, and Switzerland. Brazil meanwhile participates in the BASIC group with South Africa, India, and China, having "adopted a dual identity as a BASIC and Latin American country" (Gratius & González, 2012, p.

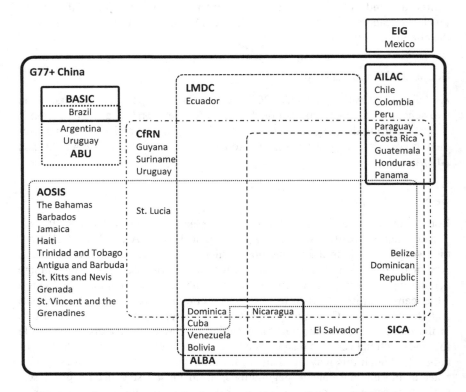

Figure 9.1 Latin American participation in negotiating coalitions at COP24.

13), reflecting Brazil's more general process of "Bricsalization" (Gratius, 2012). In more recent years Brazil has also begun to coordinate with Argentina, and Uruguay as the Group of Argentina, Brazil, and Uruguay (ABU) making a joint opening statement in the 2017 Bonn intersessional and a formal opening statement as "ABU" at COP24. This group is predominantly concerned with the role of agriculture, along with how global warming potentials are calculated for different greenhouse gases and, largely through Brazil, in rules surrounding carbon markets (Moosmann, Urrutia, Siemons, Cames, & Schneider, 2019, p. 34). Because of its more recent emergence, and because it does not seek to present as a specifically Latin American group, this chapter does not analyse ABU in any detail, but future research could examine its relationship to the two Latin American groups.

The range of coalitions in which Latin American states are active is significant. As Castro and Klöck (2020) argue, membership of multiple coalitions has the potential to either increase the support a country can draw on or, alternatively, decrease its influence if attention and resources are spread thinly across coalitions. The overlapping geometry of Latin American participation in climate negotiations certainly creates potential for these dynamics to play out, to the detriment or benefit of AILAC and ALBA.

Thematic scope

The two groups contrast sharply in thematic scope, the breadth of issues both within and beyond the climate negotiations on which the coalition is active. This is simplest for AILAC, which formed within the climate negotiations, and operates exclusively in that forum. Indeed, it is said to have been instigated by climate negotiators, rather than higher-level ministers or heads of state (Edwards, Adarve, Bustos, & Roberts, 2017).

The picture is more complex for ALBA, as ALBA membership in the climate regime is principally a function of participation in the wider ALBA-TCP economic and political regional project. Whilst for equivalent groups which exist outside of the climate negotiations, for instance the EU, (voluntary) membership of the wider project translates into (default) membership of the group within the climate change negotiations, this is not the case for ALBA.

The first spoken intervention on behalf of ALBA, by Cuba, came on the first day of the 2009 Copenhagen COP15 (ENB, 2009a). The statement identified ALBA as consisting of all nine active members of ALBA-TCP at that point,[3] and called for developed countries to "honour their climate debt", opposing attribution of responsibility towards developing countries. Since then, the coalition's membership has been in flux, evidenced by different affiliations cited during its interventions in the negotiations. Based on statements from 2012–2017 (ALBA, 2012, 2013a, b, 2014, 2015a, b, 2017) it appears that ALBA's membership in the negotiations mostly stabilised around six core members – Bolivia, Cuba, Dominica, Ecuador, Nicaragua, and Venezuela, before Ecuador exited the wider ALBA-TCP in 2018. ALBA thus constitutes a subset of the broader ALBA-TCP and no longer includes most of ALBA-TCP's small island state members (Saint Kitts and Nevis, Saint Lucia, Saint Vincent and the Grenadines, Antigua and Barbuda). The only small island states and AOSIS members operating within ALBA on climate change at the time of writing are Dominica and Cuba (a joint founder of ALBA-TCP). This reflects apparent tensions and contradictions within the climate positions and preferences of ALBA-TCP countries in the negotiations, with small island developing states with high vulnerability to climate change finding themselves grouped with Venezuela, "unmistakably a petro-state" (Edwards & Roberts, 2015a, p. 122) and, with Ecuador, a member of OPEC, a group with a longstanding reputation for obstructionism in the climate negotiations (Depledge, 2008). This may reflect a broader tendency identified by Burges (2007, p. 1353) whereby

> [c]ountries happily accept Venezuelan aid, provide support to Venezuelan causes where it is pragmatic to do so, and even adopt some elements of the Bolivarian ALBA agenda. But all this is done only when it reflects the interests of the country in question.

Given that ALBA-TCP's members have not primarily joined because of the similarities in their preferences within the climate change regime, there is clearly difficulty acting as a coalition in the climate context, an example of the "alliance dilemma" described by Van Schaik (2012, p. 6) (in respect of the G77 and EU) whereby the gains in influence from larger group membership run up against difficulty in "bring[ing] preferences into line" as larger membership becomes more diverse.

That much of ALBA-TCP's membership does not align with the group in the climate regime is perhaps also unexpected in light of the proliferation of negotiating groups which Latin American countries can join; as noted, the inactive members of ALBA within the negotiations are largely also members of the Alliance of Small Island States (AOSIS). While advocating the strongest possible action on the part of developed countries, the small island states, as members of AOSIS, have long adopted a more pragmatic and conciliatory approach to North–South politics, with its members (including ALBA-TCP member Antigua and Barbuda) active in the Cartagena Dialogue and subsequently the High Ambition Coalition. Notably, ALBA's membership almost completely overlaps with the hardline LMDCs group.

Neither AILAC nor ALBA are limited in scope to single issues within the climate negotiations; like almost all Latin American countries, all AILAC and ALBA members fall under the G77 and China (G77), a developing country group within which they have continued to operate, albeit acting as relative outliers. Both groups align explicitly with G77 positions,[4] but also demonstrate more specific policy foci which differ between the two groups. Summarising the key foci[5] of different negotiating groups in their pre-COP24 briefing for the European Parliament, Neier, Neyer, and Radunsky (2018) identify AILAC as particularly engaged on "adaption goal and communication", "scaled-up climate finance", "technology development and transfer", and "capacity building". This demonstrates no overlap with the key ALBA focuses of "ambitious mitigation action" and "loss and damage", policy focuses which are complemented by ALBA's more general emphasis on procedural justice within the climate regime detailed below.

Likewise, the policy foci identified by Neier et al. (2018) do not capture the issue with which AILAC made the most impact on the negotiations; its bridge-building approach to the Global North and willingness to see non-Annex I parties (developing countries) take on ambitious carbon-reduction targets. AILAC were distinctive at the time amongst developing country negotiating coalitions in their promotion of emission reduction commitment from all, rather than just developed Annex I countries. This was an impactful position for the group to take on its formation at COP18 in Doha because, as Depledge (2009, p. 274) argues, "The question of whether, when, and how developing countries should assume stronger commitments more comparable to those of the Annex I parties constitutes the central political dilemma of the climate change regime".

Membership

Of the dimensions which Castro and Klöck (2020) use to characterise climate coalitions, AILAC and ALBA are perhaps most alike with regards to "size", albeit

Table 9.1 Membership of AILAC and ALBA between COP21 and COP24

AILAC members	ALBA members
Chile	Bolivia
Colombia	Cuba
Costa Rica	Dominica
Guatemala	Ecuador (until 2018)
Honduras (from 2015)	Nicaragua
Panama	Venezuela
Paraguay (from 2015)	*Antigua and Barbuda
Peru	*Grenada
	*St Kitts and Nevis
	*St Lucia
	*St Vincent and the Grenadines

Note: *ALBA members are members of the wider ALBA-TCP project, but do not form an active part of the ALBA coalition in the climate negotiations.

manifesting notably different negotiating positions. AILAC and ALBA both fall below the median size for climate negotiating groups and therefore qualify as two of the 14 "small" coalitions in the UNFCCC. AILAC has the larger active membership, with eight countries participating. ALBA, as of 2019, has five members active in the negotiating coalition, a subset of the larger membership of eight countries formally participating in the ALBA-TCP project, many of whom originally participated, at least nominally, in the group's activities as a climate coalition. Recent years have seen the wider ALBA-TCP lose membership, with Honduras leaving formally in 2010 and Ecuador leaving in 2018. Over the same period AILAC's membership has grown, with Honduras joining the 2015, five years after leaving ALBA, and Paraguay also joining the group in 2015 (Edwards & Roberts, 2015c) (Table 9.1).

In their role as Latin American coalitions espousing specific positions in the UNFCCC, AILAC and ALBA draw, in ALBA's case explicitly, on well-established traditions of Latin American regional activity. These are examined below, before discussion of the two coalitions' positions within the negotiations, which occupy differing positions along the "narrative dimension of 'bridge-building' or 'upholding the North-South divide'", which Blaxekjær and Nielsen (2014, p. 761) identify within the UNFCCC.

Negotiation positions

Historical context: Latin American regionalism

The diversity of coalition membership amongst Latin American parties to the UNFCCC also echoes the historical fragmentation of regional projects on a continent where "Political unity has remained a myth and regional cooperation has not resulted in a single common project" (Bianculli, 2016, p. 155). Latin American regional discourses have broadly corresponded to two alternative visions: a Bolivarian aspiration for regional unity and autonomy, or an urge for hemispheric unity, encompassing the USA, and first articulated in the 1823 Monroe Doctrine (Bianculli, 2016; Emerson, 2014; Riggirozzi, 2012; Tussie, 2009). The origins of both visions are nearly contemporaneous; Simón Bolívar articulated his post-colonial regional project at the same time as, and partially in response to, the Monroe Doctrine, which heralded ambitions for US-led hemispheric autonomy (Puntigliano, 2013).

In the 20th and 21st centuries, Latin America has oscillated between inward and outward looking regional projects (described in Riggirozzi, 2012; Watts & Depledge, 2018). ALBA-TCP emerged in a "wave" of regionalism, defined by Riggirozzi (2012) as a "post-hegemonic" regionalism, which sought to re-focus regional projects on social factors and empowering Latin American states, frequently justified explicitly in terms of opposition to US hegemony. Founded in 2004 with Hugo Chávez and Fidel Castro as its main architects, ALBA-TCP was intended as a regional project that would resist US-led neoliberalism (Riggirozzi, 2012), in particular providing an alternative to the proposed Free Trade Area of the Americas (Burges, 2007, p. 1346), and was explicitly inspired by Bolívar's vision, with Hugo Chávez asserting that "Bolívar is a concept .... He spoke of what today we call a multipolar world" (cited in Guevara, 2005, p. 11). The ALBA-TCP project includes economic restructuring, with state-owned enterprises embracing cooperation rather than competition, and payment-in-kind bartering to maximise the social utility of interaction, in addition to more directly fostering development among its members and upholding multipolarity in global politics (Chodor & McCarthy-Jones, 2013; Tockman, 2009). Another critical driver behind the formation of ALBA-TCP was oil, with cheaper access to Venezuela's vast reserves serving as a powerful recruitment tool for other countries, notably the smaller and poorer island states (Jacóme, 2011) with Chavez's strategy described by Burges (2007, p. 1344) as one of "leverag[ing] the country's oil wealth as a device for placing Venezuela in an international leadership position, ostensibly headed towards a new, un-savage version of globalisation".

The alternative current of Latin American regionalism, projects grounded in economic liberalisation, is today manifested in the Pacific Alliance. Founded in 2012, its members, in contrast to ALBA-TCP's alternative economic experiments, "are fully committed to the rules of the game of economic globalization" (Nolte & Wehner, 2013, p. 3). This open, market-oriented regional project has seen the lifting of virtually all trade barriers between Chile, Peru, Colombia, and Mexico. Such is the ideological distinction between ALBA-TCP and the Pacific Alliance that the Alliance has been decried by ALBA-TCP heads of state and left-wing groups within Latin America as a neo-imperialist subversion of post-hegemonic regional models (Nolte & Wehner, 2013). Significantly for climate politics, the Pacific Alliance states are all members, or supporters, of AILAC,[6] while the remaining AILAC members are observers to the Alliance, with two in the process of becoming full members (Pacific Alliance, 2019). The implication appears to be that both AILAC and the Pacific Alliance are rooted in a shared regional identity, though members of both groups remain bound into other sub-regional projects within what Malamud and Gardini (2012) term the "spaghetti bowl" of Latin American regional groups.

Reflecting these complexities, it is worth noting that AILAC/Pacific Alliance regionalism differs from earlier "hemispheric unity" projects in its ambivalence towards the USA. The Pacific Alliance is oriented towards trade with emerging countries, especially Asia and China, in a context in which "AILAC countries are … demonstrating a foreign policy that is increasingly independent of the U.S." (Edwards & Roberts, 2015b). Mirroring this wider tendency, Parker, Karlsson, & Hjerpé (2015) demonstrate how Latin American climate delegates are the least likely to view the USA as a leader in the climate negotiations, whilst AILAC tends to cooperate more closely with the EU (e.g. in launching the Cartagena Dialogue) than the USA.

ALBA's negotiating positions

ALBA-TCP's Bolivarian political mission and "post-hegemonic" regionalism has translated in the climate change context into a focus on equity, climate justice, and an uncompromising interpretation of the UNFCCC principle of "common but differentiated responsibilities and respective capabilities" (CBDR-RC). ALBA emphasises the historical responsibility – or historical "debt" – of developed countries for the accumulation of carbon in the atmosphere (e.g. ALBA, 2015a, 2017; Audet, 2013; ENB, 2009a, b; Pearson, 2009). ALBA understands climate change to be "a consequence of the capitalist system, of the prolonged and unsustainable pattern of production and consumption of the developed countries, of the application and imposition of an absolutely predatory model of development on the rest of the world" (ALBA, 2009).

Proposals for developing countries to take on additional obligations, even voluntarily, have been viewed in this light as an unfair "transfer of responsibility" (Audet, 2013; ENB, 2009b). Unsurprisingly ALBA members were slow to submit their intended nationally determined contributions (INDCs) in the run-up to COP21, with only the small island state of Dominica among the 147 countries meeting the quasi-deadline of 1 October 2015. Venezuela did not submit its INDC until just after COP21, while initially Nicaragua refused to do so at all; the Nicaraguan chief negotiator justified his stance on the grounds that "We don't want to be an accomplice to taking the world to 3 to 4 degrees and the death and destruction that it represents" (cited in Pashley, 2015),[7] though Nicaragua did submit an INDC in mid-2018.

ALBA's emphasis on historical responsibility is accompanied by antagonism towards market mechanisms (such as emissions trading), which are interpreted as turning the environment into a commodity (e.g. ALBA, 2015b; Audet, 2013; Edwards & Roberts, 2015b). ALBA members, for example, were lone (and successful) hold-outs in the final stages of negotiations at COP21 against the inclusion of the term "market mechanisms" in the Paris Agreement (IETA, 2015).

In a similar vein, although the obligation on developed countries to provide full financial support to the developing world is insisted upon, financial transfers have also been characterised as "blackmail" and an attempt to pass the responsibility for mitigation onto developing countries (ENB, 2009b). When the USA allegedly threatened to withhold funding from states who did not declare voluntary commitments under the 2009 Copenhagen Accord, this was viewed from the Bolivarian perspective as a perfect storm of US hegemony and financial domination (Goldenberg, 2010; Roberts, 2011). Indeed, the financial settlement under Copenhagen was initially described by Venezuela as "imperial interests ... being imposed. We will not sell our principles ... even for 30 billion dollars"[8] (cited in Dimitrov, 2010, p. 812).

ALBA is not unique in taking these strong positions, which are shared by many other developing countries, particularly among the LMDCs (see Audet, 2013; Blaxekjær, Lahn, Nielsen, Green-Weiskel, & Fang, 2020; Dimitrov, 2010). However, ALBA distinguishes itself by occupying the more radical end of the spectrum, and framing its arguments in a broader ideological narrative. It has also sought to present an alternative framework for addressing environmental issues, based on respect for "Mother Earth" (*La Madre Tierra*; see e.g. ALBA, 2015a, b), a discourse associated in particular with Bolivia. In April 2010, Bolivia hosted a "World People's Conference on Climate Change and the Rights of Mother Earth", drawing over 30,000 participants, mostly from civil society and grassroots organizations (UN-NGLS, 2010). The outputs of the conference, including a proposed Universal Declaration of the Rights of Mother Earth, clearly posit climate change as a consequence of capitalism and its "logic of competition, progress and limitless growth" (People's Agreement, 2010), a point then taken up by ALBA in its statements on climate change (e.g. ALBA, 2015a). ALBA-TCP has also promoted the concept of Mother Earth in other international forums, such as the Convention on Biological Diversity (Borie & Hulme, 2015).

As part of its narrative of defending the oppressed against the powerful, the coalition also champions the rights of indigenous peoples (Moosmann et al., 2019, p. 34), and expresses vocal concern for civil society and youth (e.g. ALBA, 2009). The interconnections between these two causes – of indigenous peoples and the global environment – are highly potent in the Latin American context.

Another facet of ALBA's approach has been an emphasis on procedural justice, reflecting the group's self-positioning as champions of the unheard, and their willingness to be minority dissenting voices blocking a consensus.[9] Notably, ALBA's insistence on fairness, inclusivity, and transparency led four of its members[10] to block the adoption of the Copenhagen Accord, because the deal had been struck behind closed doors by a small group of "key" players (Dimitrov, 2010), with the Venezuelan delegate going so far as to describe the Copenhagen Accord as a "coup d'état against the United Nations" (Black, 2009; ENB, 2009b). A year after Copenhagen, Bolivia (not ALBA as a group) was the only country protesting the adoption of the Cancún Agreements at COP16.[11] At COP21, while not formally objecting to the Paris Agreement's adoption, Nicaragua was almost alone in powerfully decrying the treaty's weakness (UNFCCC, 2015), and then refusing to sign it.

AILAC's negotiating positions

Although formally launched in 2012, at COP18 in Doha (ENB, 2012, p. 28), AILAC's members had been coordinating their negotiating strategies for some years before (AILAC, 2017). AILAC's approach to the negotiations is explicitly conciliatory and constructive. The group presents its role as one of "build[ing] bridges between the different negotiation groups, increasing trust and broadening the space for consensus building" (AILAC, 2017), a stance that is diametrically opposed to the disruption and confrontation which has characterised ALBA.

AILAC countries are all members of the G77 group of developing countries, and coordinate closely with other members of that group on certain issues, notably finance. However, within the G77, AILAC countries overlap neither with the LMDCs, nor with AOSIS, reflecting their desire to occupy a "middle ground" in climate politics (Blaxekjær & Nielsen, 2015; Edwards et al., 2017). At the same time, AILAC's position allows for cooperation with developed states. Most of AILAC's members were also core participants in the Cartagena Dialogue, whose launch in 2010 preceded the formation of their own regional coalition based on the same principles of cooperation and pragmatism (Blaxekjær, 2020; Blaxekjær & Nielsen, 2015). Building on their involvement in the Dialogue, AILAC members were also important players in the High Ambition Coalition at COP21, and AILAC has openly cooperated with other groups pursuing "bridge-building", such as the EIG, with whom it has presented common statements, as well as the EU.

Like ALBA, AILAC pushes for an ambitious response to climate change, with developed countries in the lead, and with financial support to assist developing countries in addressing the problem. It does so, however, in a way not exclusively focused on historical wrongs, instead looking to the benefits of ambitious actions by all Parties, with a strong international regime promoting ambitious domestic policy, and vice versa. The coalition thus upholds the identity and status of its members as developing countries entitled to financial assistance, while rejecting the strictures of historical responsibility. Significantly, this approach opens the way to developing countries themselves adopting domestic mitigation strategies and international commitments, without this violating equity concerns or the CBDR-RC principle. Reflecting the deeply entrenched North/South divisions in climate politics, AILAC's alternative approach has been interpreted by some other developing countries as betrayal of developing country solidarity and CBDR-RC (Edwards et al., 2017).

AILAC has a pragmatic attitude to the role of market mechanisms in combating climate change, in line with the generally market-oriented economies of its members. AILAC members were, for example, enthusiastic participants in the clean development mechanism (CDM) under the Kyoto Protocol,[12] with all its members hosting projects, with Colombia and Peru registering more than 60 each and Chile over 100.[13] In contrast, ALBA countries have typically hosted far fewer CDM projects, with Venezuela and Dominica hosting none.

AILAC member's willingness to contribute to the regime in pragmatic ways, even if this breaks long-held conventions, has seen four – Chile, Colombia, Panama, and Peru – become contributors to the Green Climate Fund (GCF),[14] among only nine non-Annex I Parties to do so at the time of writing. Under the UNFCCC, only the most advanced economies have obligations to provide financial support to developing countries. The possibility that others might also contribute to the regime's funding mechanisms has been highly controversial, with the traditional G77 line being that this would violate the CBDR-RC principle. Mexico and South Korea (non-Annex I Parties, but not G77 members) were among the first to break from this stricture and voluntarily donate funds, and a handful of others, principally AILAC members, have followed suit.

AILAC's negotiation style stresses the pursuit of pragmatic solutions. This was exemplified in Peru's COP20 Presidency, which was applauded for quiet diplomacy, bridge-building, and commitment to securing a consensus (Edwards & Roberts, 2015b). Peru continued in this leadership role, lending unusually strong support to COP21 President Laurent Fabius, not only in the run-up to Paris, but during the conference itself, where COP20 President Manuel Pulgar-Vidal led on engagement with non-state actors and was appointed coordinator for several key issues in the last stages of the negotiations. Interestingly, Venezuela had also expressed interest in hosting COP20. The regional group GRULAC chose Peru as its nominee, while endorsing Venezuela's offer to hold a "Social Pre-COP" for civil society. Although domestic upheaval will certainly have influenced this decision, it is not difficult to imagine that Venezuela's uncompromising positions – the opposite of what is needed in a Presidency – also contributed, along with its weak domestic climate policy.

Level of formality

The final dimension by which negotiating coalitions are characterised by Castro and Klöck is that of formality: how the group is structured and how concrete its cooperation tends to be. As was the case for the membership subcategory, AILAC and ALBA share the same broad characterization as "formal" coalitions within Castro and Klöck's (2020) taxonomy, and undertake formal actions as groups e.g. putting forward submissions to the UNFCCC.

ALBA is *a priori* a formal grouping within the UNFCCC by virtue of its formal structure beyond the climate change regime. However, as noted above, many of the members of ALBA as a wider political-economic project (ALBA-TCP) do not participate in ALBA's activities within the climate change negotiations.

AILAC, though only active within the UNFCCC negotiations, demonstrates a high degree of formality, with support structures in place that allow it to function more effectively. This ranges from a group website (http://ailac.org/en/sobre/) and twitter feed (https://twitter.com/AilacCC) to formal structures which allow it to operate more effectively within the negotiations, including official daily meetings. AILAC has a secretariat, and national delegations frequently include formal AILAC representatives; for instance, the 18-strong Costa Rican delegation to the 2019 Bonn climate conference[15] contained five registered AILAC representatives. This professional presence is in part supported by funding from Germany to support AILAC as "climate leaders",[16] which includes training for AILAC negotiators.

AILAC and ALBA in the climate negotiations: engagement, activity, and cohesion

Membership engagement

In addition to considering a negotiating group's membership and size, it is important to assess its members' level of engagement in the climate negotiations. A larger number of member countries brings definite benefits, in both a strategic (bolstering claims to support and legitimacy for the group's positions) and an operational sense (increased resources or ability to block consensus), as described by Castro and Klöck (2020). The practical advantages of larger membership can be expected to accrue more to coalitions whose member countries are more engaged in the negotiations, with larger delegations. Here AILAC appears more engaged in the negotiations, with consistently larger delegations than the ALBA countries (see Figure 9.2). In the case of COP21 in Paris for instance, AILAC's total delegation size of 932 delegates dwarfed that of ALBA (131 delegates). Whilst COP21 in 2015 saw a bumper attendance by AILAC members, in particular, Peru who played a significant organizational role following COP20 in Lima (see above), AILAC delegates have consistently outnumbered their ALBA counterparts since the emergence of AILAC as a grouping in 2012.

Figure 9.2 compares AILAC members' combined delegation size with that of ALBA's in the climate negotiations (Bolivia, Cuba, Dominica, Ecuador (until COP23), Nicaragua, and Venezuela). Paraguay and Honduras are counted as part of AILAC's total from COP21, the year in which they both became formal

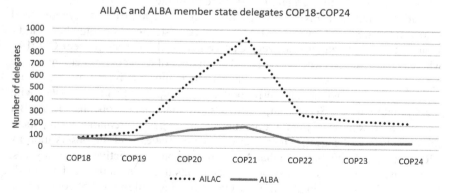

Figure 9.2 Aggregate delegation sizes for AILAC and ALBA.

members. The increase in AILAC's delegation size stands out; whilst both coalitions saw increased attendance for COP20 in Lima (potentially facilitated by the Latin American location), and COP21, this increase was far more pronounced for AILAC, and its delegations have since settled to a significantly higher level. It is worth noting here that the spike in AILAC's attendance at COP20 and COP21 is only partly explained by Peru's COP presidency; at COP21, for instance, Peru's delegation of 322 made up over a third of AILAC's total of 934, but the remaining AILAC delegations still greatly outnumbered ALBA's 181 participants. This might, *ceteris paribus*, be expected to result in greater effectiveness for AILAC, as a larger delegation size allows deeper and wider participation across the negotiations, though as noted by Martinez et al. (2019, p. 333), the overall link between delegation size and effectiveness is not clear.

Activity and cohesion

Beyond raw delegation sizes, the impact of a particular coalition can be partly understood as a result of its level of activity and cohesion; the more effective the coalition *as a coalition* the more its constituent member countries would be expected to act explicitly as part of the group, and the higher the overall level of activity associated with the group. To apply the taxonomy put forward by Castro and Klöck (2020) to this understanding, a group's level of activity and cohesion is partially shaped by its size and level of formality.

It is important to note that considering levels of activity and cohesion is an incomplete and imperfect metonym for a coalition's influence; changing agendas in the negotiations can create windows of opportunity for more effective action: for instance, when a coalition can potentially tip the balance between effective consensus or not. In addition, as Depledge (2008, p. 19) reminds us "negotiations are highly socialized and personal processes, where individual personalities and dynamics can influence outcomes as much as national positions".

Following the approach taken by Betzold, Castro, and Weiler (2012), Figure 9.3 compares the number of interventions made in COP plenaries by AILAC and ALBA as coalitions to the aggregate statements of their member states.

Data collated from ENB daily briefings from 2012 to 2018[17] and represented below in Figure 9.3 show that AILAC's collective statements outnumbered individual member statements at every Conference of the Parties (except COP19), whereas the reverse is true for ALBA. The number of collective statements made by AILAC is also increasing over time, whereas those delivered by ALBA are declining. These insights are corroborated by group statements deposited with the Climate Change Secretariat,[18] which show that, since Copenhagen, ALBA has tended to confine its joint interventions to more general remarks during opening and closing plenaries, rather than on specific issues, and the observation by Moosmann et al. (2019, p. 34) that ALBA has become less active in recent years. Of the last seven COPs, AILAC group statements have outnumbered those made by its individual states in all but one COP, indicating a high level of cohesion. In contrast, the reverse is true for ALBA, with its sole year of higher cohesion (at

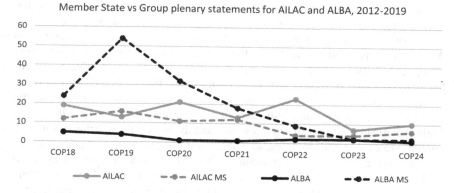

Figure 9.3 Plenary statements on behalf of AILAC/ALBA vs statements on behalf of individual member countries, COP18–COP24 (source: ENB plenary summaries, compiled by author).

COP23) appearing to be a result of the plummeting incidences of ALBA member state statements after COP19. Overall, this would suggest that AILAC is the more active and cohesive of the two negotiating coalitions.

Conclusion

The disparity in activity and cohesion between the two Latin American climate coalitions can be seen to reflect differences in the groups across the key categories by which the groups have here been compared:

Both groups fall within the cluster of regional groupings described by Castro and Klöck (2020). However, as described above, AILAC is a climate specific group, whilst ALBA in the negotiations is a subset of the wider ALBA-TCP project. AILAC's status as a climate-only coalition may encourage greater cohesion; in contrast, the full ALBA-TCP was only briefly present within ALBA in the negotiations. Whilst ALBA's existence beyond the climate negotiations means it is unlikely to disappear as a formal negotiating group, it is exposed to the fortunes of the wider ALBA-TCP project: for instance, losing Ecuador and Honduras as members.

Both groups share a clear and limited geographic scope: Latin America. Their identity and positioning in the negotiations draw on the layered history of regional groups; this is particularly explicit for ALBA. Beyond shared ground within the G77, the ideological and negotiating positions of the two groups are distinctive. This creates the possibility that ALBA's declining activity in the negotiations is at least in part related to the movement of the negotiations away from its key focuses and oppositional style. As the UNFCCC has seen the distinction between developed and developing countries erode, there may be less opportunity for ALBA to emphasise its unifying negotiation preferences. Equally, ALBA may have found itself rendered less relevant as a site for its members to direct their action and support by the overlap in its identity and positions with the LMDCs within the overlapping mosaic of Latin American participation in negotiating groups.

ALBA's relative absence of formal structures for cooperation, in contrast to AILAC, may also render its level of activity and cohesion less resilient to competing pressures and opportunities presented to its member states by alternative coalitions, and to fluctuations in the salience of key topics within the UNFCCC negotiations.

Whilst the two exclusively Latin American negotiating coalitions in the UNFCCC negotiations can be situated against the backdrop of a long history of Latin American regionalism, the pressures and dynamics of the negotiations will continue to evolve and, in doing so, shape their cohesiveness and effectiveness. This high-level overview has presented a brief description of how AILAC and ALBA fall within the taxonomy of negotiating groups proposed by Castro and Klöck (2020), and suggested the inter-relations between size, scope, and formality and the ideological and positional history to the two groups and their wider regional histories. Further research could engage more richly in the group's behaviour within the negotiations themselves, gauge the group's relative influence within the negotiations, and draw clearer links between these factors and the development of the key themes within the negotiations over time.

Notes

1 For the purposes of this chapter "ALBA" will refer to the coalition within the climate negotiations, whilst the wider regional project of the "Alianza Bolivariana para los Pueblos de Nuestra América – Tratado de Comercio de los Pueblos" (Bolivarian Alliance for the People of Our America – People's Trade Treaty) will be referred to as ALBA-TCP.
2 This chapter largely builds on a previous examination of the two groups (Watts & Depledge, 2018).
3 This excludes Honduras, which, whilst a member of ALBA at the start of COP15, announced its intention to exit the group before the conference had concluded.
4 For discussion of Latin American states' G77 identities, see Bueno Rubial (2020).
5 For which the authors drew on Earth Negotiations Bulletin summaries of recent COPs, and their own views.
6 AILAC identify Mexico and the Dominican Republic as included within the "spirit of camaraderie and inclusion" of its work; see http://ailac.org/en/sobre/.
7 Nicaragua was one of only a handful of countries that did not sign the Paris Agreement, and only announced that it would eventually join in September 2017, as a gesture of solidarity with the victims of natural disasters (Mathiesen, 2017).
8 A reference to the commitment on the part of donor countries to supply this sum as "fast-track" finance for the period 2010–2012.
9 Under the climate change regime, almost all substantive decisions must be taken by consensus, because of the absence of an agreed voting rule (Yamin & Depledge, 2004).
10 Bolivia, Cuba, Nicaragua and Venezuela, accompanied by Sudan and Tuvalu.
11 Although there are suggestions that the Bolivian chief negotiator may have gone beyond his mandate in opposing the adoption of the Agreements. He was later replaced, and the new chief negotiator made it clear that Bolivia would play a more constructive role in the future.

12 Under the CDM, developed countries (Annex I Parties) can generate credits (certified emission reductions) to offset against their emission targets, by investing in low-carbon projects in developing countries.

13 See the CDM Project Registry, https://cdm.unfccc.int/Projects/projsearch.html.

14 As of January 2019, see www.greenclimate.fund for updated figures.

15 UNFCCC 2019, "Final List of Participants", available at: https://unfccc.int/documents /197849.

16 International Climate Initiative available at: www.international-climate-initiative.com/ en/nc/de tails/?projectid=331&cHash=15a00dcc227d1a91d5807c9cb2afa204.

17 Available at www.iisd.ca

18 Available through the search engine at www.unfccc.int.

References

AILAC. (2017). AILAC website. Retrieved from http://ailac.org/en/sobre/

ALBA. (2009). ALBA declaration on Copenhagen climate summit. Retrieved from https:/ /venezuelanalysis.com/analysis/5038

ALBA. (2012, November 26). Discurso de apertura de la COP-18.

ALBA. (2013a, November 12). Intervención del ALBA. Plenaria de apertura del Grupo de Trabajo de la Plataforma de Durban.

ALBA. (2013b, June 13). Statement ALBA. Plenaria de cierre para ADP 2/2.

ALBA. (2014, December 1). Intervención del ALBA Sesión de apertura COP-20/ CMP-10.

ALBA. (2015a, October 19). Estado Plurinacional de Bolivia en Nombre de la Alianza Bolivariana para les Pueblos de Nuestra America—ALBA. UNFCCC-ADP 2/11.

ALBA. (2015b, June 1). Intervencion de la Republica Bolivariana de Venezuela en nombre de mimebros de la Alianza Bolivariana para los Pueblos de Nuestra America.

ALBA. (2017, November 6). Statement by ALBA at the opening ceremony of COP 23 of the UNFCCC, Bonn.

Audet, R. (2013). Climate justice and bargaining coalitions: A discourse analysis. *International Environmental Agreements: Politics, Law and Economics, 13*, 369–386.

Betzold, C., Castro, P., & Weiler, F. (2012). AOSIS in the UNFCCC negotiations: From unity to fragmentation? *Climate Policy, 12*(5), 591–613.

Bianculli, A. (2016). Latin America. In T. Börzel & T. Risse (Eds.), *The Oxford Handbook of Comparative Regionalism* (pp. 154–175). Oxford: Oxford University Press.

Black, R. (2009, December 19). Key powers reach compromise at climate summit. *BBC*. Retrieved from http://news.bbc.co.uk/1/hi/world/europe/8421935.stm

Blaxekjær, L. Ø. (2020). Diplomatic Learning and Trust: How the Cartagena Dialogue Brought UN Climate Negotiations Back on Track. In C. Klöck, P. Castro, F. Weiler, & L. Ø. Blaxekjær (Eds.), *Coalitions in the Climate Change Negotiations*. Abingdon: Routledge.

Blaxekjær, L. Ø., Lahn, B., Nielsen, T. D., Green-Weiskel, L., & Fang, F. (2020). The Narrative Position of the Like-Minded Developing Countries in Global Climate Negotiations. In C. Klöck, P. Castro, F. Weiler, & L. Ø. Blaxekjær (Eds.), *Coalitions in the Climate Negotiations*. Abingdon: Routledge.

Blaxekjær, L. Ø., & Nielsen, T. D. (2015). Mapping the narrative positions of new political groups under the UNFCCC. *Climate Policy, 15*(6), 751–766.

Borie, M., & Hulme, M. (2015). Framing global biodiversity: IPBES between mother earth and ecosystem services. *Environmental Science and Policy, 54*, 487–496.

Bueno Rubial, M. (2020). Identity-Based Cooperation in the Multilateral Negotiations on Climate Change: The Group of 77 and China. In C. Lorenzo (Ed.), *Latin America in Times of Global Environmental Change* (pp. 57–74). Cham: Springer.

Burges, S. W. (2007). Building a global southern coalition: The competing approaches of Brazil's Lula and Venezuela's Chávez. *Third World Quarterly, 28*(7), 1343–1358.

Castro, P., & Klöck, C. (2020). Fragmentation in the Climate Change Negotiations: Taking Stock of the Evolving Coalition Dynamics. In C. Klöck, P. Castro, F. Weiler, & L. Ø. Blaxekjær (Eds.), *Coalitions in the Climate Change Negotiations*. Abingdon: Routledge.

Chodor, T., & McCarthy-Jones, A. (2013). Post-liberal regionalism in Latin America and the influence of Hugo Chávez. *Journal of Iberian and Latin American Research, 19*(2), 211–223.

Depledge, J. (2008). Striving for no: Saudi Arabia in the climate change regime. *Global Environmental Politics, 8*(4), 9–35.

Depledge, J. (2009). The road less travelled: Difficulties in moving between annexes in the climate change regime. *Climate Policy, 9*(3), 273–287.

Dimitrov, R. (2010). Inside UN climate change negotiations: The Copenhagen conference. *Review of Policy Research, 27*(6), 795–821.

Edwards, G., Adarve, I. C., Bustos, M. C., & Roberts, J. T. (2017). Small group, big impact: How AILAC helped shape the Paris Agreement. *Climate Policy, 17*(1), 71–85.

Edwards, G., & Roberts, J. T. (2015a). *Fragmented Continent: Latin America and the Global Politics of Climate Change*. Cambridge, MA: MIT Press.

Edwards, G., & Roberts, J. T. (2015b). Latin America and UN climate talks: Not in harmony. *Americas Quarterly*. Retrieved from https://www.americasquarterly.org/content/latin-america-and-un-climate-talks-not-harmony

Edwards, G., & Roberts, J. T. (2015c). Paraguay's surprisingly powerful voice in climate negotiations. *Americas Quarterly*. Available at: https://www.americasquarterly.org/article/paraguays-surprisingly-powerful-voice-in-climate-negotiations/

Emerson, R. G. (2014). An art of the region: Towards a politics of regionness. *New Political Economy, 19*(4), 559–577.

ENB. (2009a). Copenhagen highlights: Monday, 7 December 2009. *Earth Negotiations Bulletin, 12*(449), 1–4. Available at: https://enb.iisd.org/download/pdf/enb12449e.pdf.

ENB. (2009b). Summary of the Copenhagen climate conference. *Earth Negotiations Bulletin, 12*(459). 1-30. Available at: https://enb.iisd.org/download/pdf/enb12459e.pdf.

ENB. (2012). Summary of the Doha climate change conference: 26 November–8 December 2012. *Earth Negotiations Bulletin, 12*(567), 1–30. Available at: https://enb.iisd.org/download/pdf/enb12567e.pdf.

Goldenberg, S. (2010, April 9). US denies climate aid to countries opposing Copenhagen accord. *The Guardian*.

Gratius, S. (2012). Brazil and the European Union: Between Balancing and Bandwagoning. In European Strategic Partnerships Observatory Working Paper 2/12. Fundación para las Relaciones Internacionales y el Diálogo Exterior (FRIDE).

Gratius, S., & González, D. (2012). The EU and Brazil: Shared Goals, Different Strategies. In G. Grevi & T. Renard (Eds.), *Hot Issues, Cold Shoulders, Warm Partners: EU Strategic Partnerships and Climate Change* (pp. 11–22). Belgium: Egmont Institute.

Guevara, A. (2005). *Chávez: Venezuela and the New Latin America—An Interview with Hugo Chávez*. Melbourne: Ocean Press.

IETA. (2015, December 14). IETA Paris COP 21 summary: The makings of a global climate deal—And a new era for carbon markets. Retrieved from http://www.ieta.org/resources/Conferences_Events/COP21/IETA_COP_21_Summary.pdf.

Jacóme, F. (2011). *Petrocaribe: The Current Phase of Venezuela's Oil Diplomacy in the Caribbean*. Policy paper 40, Friedrich Ebert Siftung, Programa de Cooperación en Seguridad Regional. Available from: http://library.fes.de/pdf-files/bueros/la-seguridad/08723.pdf.

Malamud, A., & Gardini, G. L. (2012). Has regionalism peaked? The Latin American quagmire and its lessons. *The International Spectator, 47*(1), 116–133.

Martinez, G. S., Hansen, J. I., Olsen, K. H., Ackom, E. K., Haselip, J. A., von Kursk, O. B., & Dunbar, M. B.-N. (2019). Delegation size and equity in climate negotiations: An exploration of key issues. *Carbon Management, 10*(4), 431–435.

Mathiesen, K. (2017, September 21). Nicaragua to join Paris climate deal in solidarity with 'first victims'. *Climate Home News*. Retrieved from https://www.climatechangenews.com/2017/09/21/nicaragua-join-paris-climate-deal-solidarity-first-victims/.

Moosmann, L., Urrutia, C., Siemons, A., Cames, M., & Schneider, L. (2019). *International Climate Negotiations: Issues at Stake in View of the COP 25 UN Climate Change Conference in Madrid*. Retrieved Brussels from https://www.europarl.europa.eu/RegData/etudes/STUD/2019/642344/IPOL_STU%282019%29642344_EN.pdf.

Neier, H., Neyer, J., & Radunsky, K. (2018). *International Climate Negotiations: Issues at Stake in View of the COP 24 UN Climate Change Conference in Katowice and Beyond*. Retrieved Brussels from http://www.europarl.europa.eu/RegData/etudes/STUD/2018/626092/IPOL_STU(2018)626092_EN.pdf

Nolte, D., & Wehner, L. (2013). The Pacific Alliance Casts Its Cloud Over Latin America. In *GIGA Focus International Edition No. 8*. Hamburg: GIGA German Institute of Global and Area Studies.

Pacific Alliance. (2019). Observer states to the Pacific alliance. Retrieved from https://alianzapacifico.net/en/observant-countries/

Parker, C. F., Karlsson, C., & Hjerpé, M. (2015). Climate change leaders and followers: Leadership recognition and selection in the UNFCCC negotiations. *International Relations, 29*(4), 434–454.

Pashley, A. (2015). Nicaragua to defy UN in climate pledge refusal. *Climate Change News*. Retrieved from http://www.climatechangenews.com/2015/11/30/nicaragua-to-defy-un-in-climate-pledge-refusal/

Pearson, T. (2009). ALBA declaration on Copenhagen climate summit.

People's Agreement. (2010). World people's conference on climate change and the rights of mother earth. Retrieved from https://pwccc.wordpress.com/support/

Puntigliano, A. R. (2013). Geopolitics and Integration: A South American Perspective. In A. R. Puntigliano & J. Briceño-Ruiz (Eds.), *Resilience of Regionalism in Latin America and the Caribbean* (pp. 19–52). Basingstoke: Palgrave Macmillan.

Riggirozzi, P. (2012). Region, regionness and regionalism in Latin America: Towards a new synthesis. *New Political Economy, 17*(4), 421–443.

Roberts, J. T. (2011). Multipolarity and the new world (dis)order: US hegemonic decline and the fragmentation of the global climate regime. *Global Environmental Change, 21*, 776–784.

Tockman, J. (2009). The rise of the "Pink Tide": Trade, integration and economic crisis in Latin America. *Georgetown Journal of International Affairs, 10*(2), 31–39.

Tussie, D. (2009). Latin America: Contrasting motivations for regional projects. *Review of International Studies, 35*, 169–188.

UN-NGLS. (2010). World people's conference on climate change and the rights of mother earth. Retrieved from https://unngls.org/index.php/un-ngls_news_archives/2010/792-world-people%E2%80%99s-conference-on-climate-change-and-the-rights-of-mother-earth

UNFCCC. (2015). On demand video, conference of the parties 11th meeting. Retrieved from http://unfccc6.meta-fusion.com/cop21/events/2015-12-12-17-26-conference-of-the-parties-cop-11th-meeting

Van Schaik, L. (2012). The EU and the Progressive Alliance Negotiating in Durban: Saving the Climate? In ODI Working Paper 354. London: Overseas Development Institute.

Watts, J., & Depledge, J. (2018). Latin America in the climate change negotiations: Exploring the AILAC and ALBA coalitions. *Wiley Interdisciplinary Reviews: Climate Change*, 9(6), e533.

Yamin, F., & Depledge, J. (2004). *The International Climate Change Regime: A Guide to Rules, Institutions and Procedures*. Cambridge: Cambridge University Press.

Part 3

Conclusion and Outlook

Part 7

Conclusion and Outlook

10 Conclusions

Florian Weiler, Paula Castro, and Carola Klöck

We started this volume by noting the fundamental role of coalitions in structuring and shaping multi-party, multi-issue negotiations, such as those on climate change. Coalitions allow parties with some common objectives to share power and resources to reach those objectives (Narlikar, 2003; Rubin & Zartman, 2000), while also reducing the complexity of the negotiations (Dupont, 1996), thus making the process more manageable and increasing the chance of a positive outcome. In short, understanding coalitions and coalition dynamics is essential to understanding international (climate) negotiations.

Yet, despite the central role of coalitions – defined here as repeated cooperative efforts between at least two parties to obtain common goals – we still know comparatively little about coalition formation, maintenance, or effectiveness (Blaxekjær & Nielsen, 2015; Carter, 2015; Drahos, 2003; Gray, 2011). Much coalition research remains conceptual in nature. Even applications to the climate change negotiations are often of a (game) theoretical nature (e.g. Buchner & Carraro, 2009; Woods & Kristófersson, 2016; Wu & Thill, 2018). Other works take a more descriptive approach and focus on the first-hand experiences of negotiators (Bueno Rubial & Siegele, 2020).

This book extends this research by addressing some of the gaps in coalition and climate negotiations scholarship. More specifically, we have two objectives: first, we want to take stock of research on coalitions in the climate change negotiations, but employing a more systematic and comprehensive approach, and second, we seek to contribute to this body of work both from a theoretical and empirical perspective. We do so by examining a plurality of coalitions, as well as exploring developments over time.

As there is still "a lacuna in the literature on coalition-building and coalition diplomacy in the [climate] regime more broadly" (Carter, 2015, p. 217), the first part of this book is concerned with generally taking stock of the various climate coalition groups, and more specifically, the patterns of coalition formation and maintenance. Castro and Klöck (2020; Chapter 2) identify three clusters or coalition types; Weiler and Castro (2020; Chapter 3) relate coalition type to coalition behaviour and centrality in the negotiations; and Chan (2020; Chapter 4) situates the emergence of subgroups within the broad group of developing countries, the Group of 77 and China (G77) in the historical evolution of the climate negotiations.

The second part then examines a number of individual coalitions, which are very diverse in terms of their history, modus operandi, positions, and role in the negotiations. The Pacific small island states have turned out to be pivotal players since the very start of the climate negotiations (Carter, 2020; Chapter 5). The Cartagena Dialogue on Progressive Action served as an informal platform to bring the negotiations back on track after the 2009 Copenhagen Summit (Blaxekjær, 2020; Chapter 6). The Like-Minded Developing Countries (LMDCs) focus on the differential treatment of developed (Annex I) and developing (non-Annex I) countries in a post-Kyoto climate regime (Blaxekjær, Lahn, Nielsen, Green-Weiskel, & Fang, 2020; Chapter 7). The African Group of Negotiators (AGN) is the only regional UN group active in substantive negotiations but sometimes struggles to present one common voice (Chin-Yee, Nielsen, & Blaxekjær, 2020; Chapter 8). Finally, Latin America has not produced one common voice, but two opposing approaches to the negotiations, embodied in the more radical Bolivarian Alliance of the Peoples of Our America (ALBA) on the one hand, and the more conciliatory Independent Association of Latin American and the Caribbean (AILAC) on the other (Watts, 2020; Chapter 9).

What have we learnt from these various contributions? In this concluding chapter, we discuss some recurring themes and common findings that emerge across the diverse coalitions and chapters of this volume. Finally, we also discuss open questions that still need to be addressed, and ways forward for (climate) coalition analysis and research.

Recurring themes and results

Despite their diversity in focus and methods, the chapters in this volume yield some common themes and results. In particular, we here identify four core findings: first, coalitions are context-specific; time plays an important role. Second, coalitions are "sticky" and tend to persist over time, although their level of activity and influence may change. Third, there seems to be a hierarchy of coalitions, as coalitions operate at different levels. Fourth, given the proliferation of coalitions, and the different "layers", overlapping memberships are inevitable. Most countries belong to more than one coalition in the climate change negotiations, which can have both positive and negative effects on their participation and influence. Let us discuss these four points in more detail.

A first key result pertains to the context specificity of coalitions. Coalitions must be studied and understood in the context of the overall negotiations. Coalition emergence is closely linked to overall negotiation dynamics, as Chan (2020) describes in detail. Interestingly, the temporal dimension is important for both regional and climate-specific coalitions. For climate-specific coalitions, the link is rather unsurprising: they were created specifically in the context of the climate change negotiations, at a specific point in time in the negotiations.

For example, the Cartagena Dialogue was a direct response to the failure of the 2009 Copenhagen Summit. Without the Copenhagen fiasco, the Cartagena Dialogue would simply not have been set up, nor would it have taken its specific informal format (Blaxekjær, 2020). Similarly, the emergence of the LMDCs can be understood as a direct response to the discussions around how developed and developing countries should be differentiated – or not – in the 2015 Paris Agreement (Blaxekjær et al., 2020). In both cases, however, the creation of a new coalition built on previous contacts and longer-term instances of exchange and collaboration. Yet even groups that have their roots outside of the climate context, such as ALBA, become active in the climate negotiations as a result of certain developments and dynamics in the negotiations. ALBA was originally a regional trade agreement, but also started to coordinate within the climate change negotiations around the 2009 Copenhagen Accord, when market mechanisms – opposed by ALBA – took centre stage (Watts, 2020). The AGN has existed as a regional UN group since before the start of the negotiations, but only became more active and vocal over time, as core issues for Africa became more prominent on the climate agenda, notably adaptation, loss and damage, and adaptation finance (Chin-Yee et al., 2020). In other words, there is a two-way relationship between coalitions and the overall negotiations; coalitions shape overall negotiation dynamics – and overall negotiation dynamics shape coalitions.

Second, in contrast to some prior research (Chasek, 2005; Drahos, 2003), we find that coalitions tend to be "sticky" and persist over time. Even if coalitions emerge at a specific point in time, to achieve a specific objective, they tend to persist even when this objective has been met (or not), rather than disband (Castro & Klöck, 2020). For example, the LMDCs sought to uphold the strict division of countries into developed (Annex I) and developing (non-Annex I) countries, with binding greenhouse gas emission reductions only for the former. The Paris Agreement instead treats all countries in the same way and requires nationally determined contributions (NDCs) from *all* parties. Although the original purpose of the LMDCs is thus no longer relevant, the LMDCs continue to meet and coordinate (Blaxekjær et al., 2020). Coalition formation seems to be rather resource intensive (see also Narlikar, 2003), and members are reluctant to give up a platform once it exists. Yet, persistence does not imply a constant level of activity or influence, which can and do fluctuate over time (Weiler & Castro, 2020). There are periods in the negotiations in which a coalition is more active and vocal – such as ALBA prior to the 2009 Copenhagen Summit – and other periods in which the coalition mainly seems to exist on paper (Watts, 2020). But coalitions, once formed, can be "resurrected": the High Ambition Coalition (HAC), for example, was forged to make the 2015 Paris Agreement possible (Mathiesen & Harvey, 2015), and reappeared at subsequent COPs, when negotiations again threatened to stall (Sutter, 2018).

Third, there seems to be a hierarchy of coalitions. Coalitions operate at different levels. Some coalitions could be described as "core coalitions"; these are mainly the coalitions that have existed since the early years of the climate change negotiations, such as the Alliance of Small Island States (AOSIS) (Carter, 2020) and, above all, the G77 (Chan, 2020). These core coalitions tend to be global in geographic scope, have a large membership, and have more formalised structures. They also tend to be thematically rather broad – their focus tends to cover a wide range of issues in the negotiations. Even though AOSIS emerged as a climate-specific coalition as a result of its members' stark vulnerability to climate change, it has grown to issue opinions and become influential across many aspects of the climate negotiations and beyond (Carter, 2020). While having rather formal structures enhances the cohesion of these core coalitions and thus increases the likelihood that they persist and remain relevant over time, their global nature and large size tend to challenge such cohesiveness. As the range of issue areas addressed by the climate negotiations has grown over time, and as the members of these core coalitions have evolved economically, differences in interests and priorities among them have emerged. This has been clearly shown by the literature for the case of the G77 (see Chan, 2020; Kasa, Gullberg, & Heggelund, 2008; Vihma, Mulugetta, & Karlsson-Vinkhuyzen, 2011), but also discussed to a certain extent for AOSIS (Betzold, Castro, & Weiler, 2012). One strategy to cope with this fragmentation has been to allow the emergence of subgroups or sub-coalitions focused on more specific issue areas. While the core coalition keeps being active around the topics on which its members can agree common positions, and, crucially, as a source of common identity and a venue for exchange and institutional support (Carter, 2020; Chan, 2020), the sub-coalitions allow parties the flexibility to pursue their own specific interests and positions. Many of the newer regional or climate-specific coalitions that have emerged in later rounds of negotiations, and in particular around the 2009 Copenhagen and 2015 Paris Summits are such sub-coalitions, and more or less explicitly understand themselves as such. The LMDCs, for example, very explicitly position themselves as anchored in the G77 and as "true" representatives of the Global South (Blaxekjær et al., 2020), while the Pacific small island developing states (Pacific SIDS) started to coordinate to have a Pacific voice within AOSIS, which remains the "core" coalition for island states (Carter, 2020). Sub-coalitions thus not only seek to affect the overall negotiations, but also shape negotiations *within* broader coalitions, *within* G77 or *within* AOSIS. At the same time, there are "meta-coalitions" like the Cartagena Dialogue or HAC that represent a "coalition of coalitions". Here, the idea is to bring together like-minded negotiators from across the diverse (core) coalitions and to bridge divides. The meta-coalitions serve as platforms for exchange and dialogue between coalitions and across divides, notably the divide between the developed North and the developing South. Coalitions thus add not only one meso-layer to multilateral negotiations, but several layers: we have individual countries, sub-coalitions, core coalitions, meta-coalitions, and finally, plenary negotiations.

If we understand coalitions as operating at different levels, we also better understand the proliferation of different and diverse coalitions that have changed the overall negotiation dynamics. This proliferation of diverse coalitions also means that, fourth, and finally, we find overlapping memberships; countries routinely belong to more than one coalition, and may – at least formally – be a member of five, six, or more different coalitions. The result is a complex web of partially overlapping coalitions (Castro & Klöck, 2020). How do these overlapping, multiple coalition affiliations influence the bargaining power and negotiation capacity of individual countries, in particular the smaller and poorer ones that depend disproportionately on cooperation and coalitions? Castro and Klöck (2020) expected both positive and negative effects, and this seems to be the case in practice. For example, the Pacific SIDS seemed to have benefitted from their ties to several coalitions. Multiple coalition affiliations helped Pacific Islands to reach out to other coalitions, identify common objectives, and create partnerships. The Pacific position was thus represented in multiple forums (Carter, 2020). The Cartagena Dialogue, but also AILAC, similarly sought to build bridges and create new partnerships (Blaxekjær, 2020; Watts, 2020). In contrast, multiple coalition affiliations seem to have had mainly negative effects for the AGN. Overlapping memberships – in part with coalitions that defend very different objectives, such as OPEC – weakened the strategy of "One Africa, One Voice" and created tensions rather than synergies (Chin-Yee et al., 2020).

Ways forward for coalition research

Overall, we hope that the contributions in this volume highlight the diversity of coalitions and forms of cooperation in multilateral (climate) negotiations, and that this book adds to our still patchy understanding of coalition dynamics. The authors of this book present new ways to categorise the growing number of coalitions and demonstrate that this new classification scheme is useful, as well as analyse some coalitions in more detail through case studies. Yet, many questions remain open. Here, we list some of these open questions, and discuss some potential ways forward for coalition research.

First, we have not assessed here in a systematic manner how successful individual countries and coalitions are, and why. This seems one important way forward. The literature suggests that the behaviour and the success of coalitions depends on the decision-making rules in place (Odell, 2013; Panke, 2012). When decisions are made by consensus, as in the climate negotiations, one of the most effective tactics for coalitions is the threat to block consensus. This threat is more credible the more homogeneous the coalition members' preferences are, the more powerful players the coalition includes, and the larger the coalition is (see e.g. Ciplet, Khan, & Roberts, 2015; Narlikar & Odell, 2006; Odell & Sell, 2006; Panke, 2012). In the climate negotiations, we see fragmentation into more and more coalitions, often smaller sub-coalitions. This fragmentation on the one hand increases homogeneity, yet decreases the power potential of these coalitions on the other. How do countries and coalitions understand and respond to this dilemma? Do we indeed see a relationship between size, homogeneity, and influence? Both qualitative case studies and quantitative analyses would be useful to address these questions. Success is difficult to measure and track, in particular in settings as complex as the climate change negotiations. It is thus important to consider negotiators' perspectives and their perceived influence. At the same time, quantitative approaches, such as network analyses, can identify patterns across time and a large number of actors and coalitions. Recent developments in network analysis (see e.g. Krivitsky & Handcock, 2016; Leifeld, Cranmer, & Desmarais, 2018) are promising in this respect.

A second question directly results from the previous observation that the number of coalition groups is growing, but focuses on a different level of analysis, that of the individual country. As noted above, the proliferation of coalitions leads to countries being a member in two, three, or more coalitions. In particular, smaller countries that have fewer resources tend to join more coalitions (Castro & Klöck, 2018; Ciplet et al., 2015). We already noted that multiple coalition affiliations may increase or decrease individual countries' influence (Castro & Klöck, 2020; see also Klöck, 2020). But multiple coalition affiliations also pose very practical questions: how do countries – particularly those with very small delegations of just two or three negotiators – handle and manage their multiple coalition memberships? Do they focus their activities on one (or several) coalition(s) of particular importance to them, and if so, how do they decide which coalition is important? Or do they try to use their multiple memberships simultaneously to gain influence? A related question concerns the very different coalition strategies of individual countries. While some countries participate in a variety of coalitions, there are also countries that belong to only one coalition (notably many of the developed countries), as well as some countries that are not part of any coalition (see Figure 1.1 in Klöck, Castro, Weiler, & Blaxekjær, 2020; see also Appendix I). The questions still unanswered are therefore which countries join coalitions, which ones, and why? Why do some countries belong to multiple coalitions, and others to none? How are these decisions made?

Third, we here examine only a subset of all the different coalitions active in the climate negotiations. Based on the negotiation summaries of the Earth Negotiation Bulletin (International Institute for Sustainable Development, 1997–2018), Klöck and Castro (2020) identify 25 different coalitions (for an overview of these coalitions, see also the Appendix II). In addition, there are more short-lived instances of cooperation such as the Green Group at COP1 (Yamin & Depledge, 2004). While some coalitions, like the G77, BASIC (Brasil-South Africa-India-China), or the European Union, have received considerable academic (and political attention), surprisingly little has been written on other coalitions, such as the Commission des Forêts d'Afrique Centrale (COMIFAC), the Central Asia, Caucasus, Albania, and Moldova Group (CACAM), or the Environmental Integrity Group (EIG). More work on these less well-researched coalitions would be very welcome. Similarly, we focus here on coalitions in the climate negotiations – but there are many more multilateral negotiations where coalitions play a central role. As noted at the outset, coalitions in the trade negotiations have received some attention (e.g. Cepaluni, Galdino, & de Oliveira, 2012; Narlikar, 2003), as well as those in the UN General Assembly (Laatikainen & Smith, 2020b), whereas less has been written for example about the biodiversity or desertification negotiations, or other arenas. Here, comparative research seems particularly fruitful, not least because this would allow the teasing out of the role of context factors, such as the decision rule mentioned earlier (Odell, 2013).

Concluding remarks

Coalitions are an elemental part of the international climate negotiations, a component without which many developments of the UNFCCC negotiation process since its inception cannot be fully understood. It is all the more surprising that our knowledge of how these various coalitions operate, and how they influence the decision-making process during the negotiations, is still so limited. To a large degree, this is of course related to one of the points we raised above in this concluding chapter, i.e., that many countries and parties are members of multiple coalitions, forming a complex and confusing web of overlapping coalitions, which significantly increases the difficulty of systematic empirical analysis. However, it is our view that in order to understand the collective action process represented by the climate negotiations, we need a more complete picture of the role coalitions play in aggregating the interests of the individual members states to find solutions all can agree on. In this process, coalitions play an important bridging role between the national and the international level, and could even be understood as a third, meso-level in the complex system of climate governance. We agree strongly with Laatikainen and Smith (2020a, p. 318) who, writing on the UN system in general, note that "multilateral diplomacy is inseparable from group relations … because even though member states individually exercise permanent representation, it is the nature of multilateralism to produce collectively determined outcomes". We hope the first part of this book is a stepping stone towards a better understanding of this bridging function played by coalition groups in general, and that the chapters in Part II shed light on the how individual coalitions function, how they have played their part in finding common ground, and how they have so far shaped the story of the climate change negotiations.

References

Betzold, C., Castro, P., & Weiler, F. (2012). AOSIS in the UNFCCC negotiations: From unity to fragmentation? *Climate Policy*, *12*(5), 591–613.

Blaxekjær, L. Ø. (2020). Diplomatic Learning and Trust: How the Cartagena Dialogue Brought UN Climate Negotiations Back on Track. In C. Klöck, P. Castro, F. Weiler, & L. Ø. Blaxekjær (Eds.), *Coalitions in the Climate Change Negotiations*. Abingdon: Routledge.

Blaxekjær, L. Ø., Lahn, B., Nielsen, T. D., Green-Weiskel, L., & Fang, F. (2020). The Narrative Position of the Like-Minded Developing Countries in Global Climate Negotiations. In C. Klöck, P. Castro, F. Weiler, & L. Ø. Blaxekjær (Eds.), *Coalitions in the Climate Negotiations*. Abingdon: Routledge.

Blaxekjær, L. Ø., & Nielsen, T. D. (2015). Mapping the narrative positions of new political groups under the UNFCCC. *Climate Policy*, *15*(6), 751–766.

Buchner, B., & Carraro, C. (2009). Parallel Climate Blocs, Incentives to Cooperation in International Climate Negotiations. In R. Guesnerie & H. Tulkens (Eds.), *The Design of Climate Policy* (pp. 137–163). Cambridge, MA: MIT Press.

Bueno Rubial, M., & Siegele, L. (Eds.). (2020). *Negotiating Climate Change Adaptation: The Common Position of the Group of 77 and China*. Cham: Springer.

Carter, G. (2015). Establishing a Pacific Voice in the Climate Change Negotiations. In G. Fry & S. Tarte (Eds.), *The New Pacific Diplomacy* (pp. 205–220). Canberra: ANU Press.

Carter, G. (2020). Pacific Island States and 30 Years of Global Climate Change Negotiations. In C. Klöck, P. Castro, F. Weiler, & L. Ø. Blaxekjær (Eds.), *Coalitions in the Climate Negotiations*. Abingdon: Routledge.

Castro, P., & Klöck, C. (2018). Coalitions in Global Climate Change Negotiations. In *INOGOV Policy Brief N°5*. INOGOV: Innovations in Climate Governance.

Castro, P., & Klöck, C. (2020). Fragmentation in the Climate Change Negotiations: Taking Stock of the Evolving Coalition Dynamics. In C. Klöck, P. Castro, F. Weiler, & L. Ø. Blaxekjær (Eds.), *Coalitions in the Climate Change Negotiations*. Abingdon: Routledge.

Cepaluni, G., Galdino, M., & de Oliveira, A. J. (2012). The bigger, the better: Coalitions in the GATT/WTO. *Brazilian Political Science Review, 6*(2), 28–55.

Chan, N. (2020). The Temporal Emergence of Developing Country Coalitions. In C. Klöck, P. Castro, F. Weiler, & L. Ø. Blaxekjær (Eds.), *Coalitions in the Climate Change Negotiations*. Abingdon: Routledge.

Chasek, P. S. (2005). Margins of power: Coalition building and coalition maintenance of the South Pacific island states and the alliance of small island states. *Review of European Comparative and International Environmental Law, 14*(2), 125–137.

Chin-Yee, S., Nielsen, T. D., & Blaxekjær, L. Ø. (2020). One Voice, One Africa: The African Group of Negotiators—Regional Drivers of Climate Policy. In C. Klöck, P. Castro, F. Weiler, & L. Ø. Blaxekjær (Eds.), *Coalitions in the Climate Negotiations*. Abingdon: Routledge.

Ciplet, D., Khan, M., & Roberts, J. T. (2015). *Power in a Warming World: The New Global Politics of Climate Change and the Remaking of Environmental Inequality*. Boston: MIT Press.

Drahos, P. (2003). When the weak bargain with the strong: Negotiations in the World Trade Organization. *International Negotiation, 8*, 79–109.

Dupont, C. (1996). Negotiation as coalition building. *International Negotiation, 1*(1), 47–64.

Gray, B. (2011). The complexity of multiparty negotiations: Wading into the muck. *Negotiation and Conflict Management Research, 4*(3), 169–177.

International Institute for Sustainable Development. (1997–2018). Earth negotiation bulletins. Retrieved from http://enb.iisd.org/enb

Kasa, S., Gullberg, A. T., & Heggelund, G. (2008). The group of 77 in the international climate negotiations: Recent developments and future directions. *International Environmental Agreements: Politics, Law and Economics, 8*(2), 113–127.

Klöck, C. (2020). Multiple coalition memberships: Helping or hindering small states in multilateral (climate) negotiations? *International Negotiation, 25*(2), 279–297.

Klöck, C., & Castro, P. (2020). Fragmentation in the Climate Change Negotiations: Taking Stock of the Evolving Coalition Dynamics. In C. Klöck, P. Castro, F. Weiler, & L. Ø. Blaxekjær (Eds.), *Coalitions in the Climate Change Negotiations*. Abingdon: Routledge.

Klöck, C., Castro, P., Weiler, F., & Blaxekjær, L. Ø. (2020). Introduction. In C. Klöck, P. Castro, F. Weiler, & L. Ø. Blaxekjær (Eds.), *Coalitions in the Climate Negotiations*. Abingdon: Routledge.

Krivitsky, P. N., & Handcock, M. (2016). Tergm: Fit, simulate and diagnose models for network evolution based on exponential-family random graph models. *Statnet Project.* Retrieved from http://www.statnet.org. *R package version, 3(0)*.

Laatikainen, K. V., & Smith, K. E. (2020a). Conclusion: "The Only Sin at the UN Is Being Isolated". In K. E. Smith & K. V. Laatikainen (Eds.), *Group Politics in UN Multilateralism* (pp. 305–322). Leiden and Boston: Brill Nijhoff.

Laatikainen, K. V., & Smith, K. E. (2020b). Introduction: Group Politics in UN Multilateralism. In K. E. Smith & K. V. Laatikainen (Eds.), *Group Politics in UN Multilateralism* (pp. 3–19). Leiden and Boston: Brill Nijhoff.

Leifeld, P., Cranmer, S. J., & Desmarais, B. A. (2018). Temporal exponential random graph models with btergm: Estimation and bootstrap confidence intervals. *Journal of Statistical Software, 83*(6), 1–36.

Mathiesen, K., & Harvey, F. (2015, December 8). Climate coalition breaks cover in Paris to push for binding and ambitious deal. *The Guardian*. Retrieved from https://www.the guardian.com/environment/2015/dec/08/coalition-paris-push-for-binding-ambitious-climate-change-deal

Narlikar, A. (2003). *International Trade and Developing Countries: Bargaining Coalitions in the GATT & WTO*. London and New York: Routledge.

Narlikar, A., & Odell, J. S. (2006). The Strict Distributive Strategy for a Bargaining Coalition: The Like-Minded Group in the World Trade Organization. In J. S. Odell (Ed.), *Negotiating Trade: Developing Countries in the WTO and NAFTA* (pp. 115–144). Cambridge: Cambridge University Press.

Odell, J. S. (2013). Negotiation and Bargaining. In W. Carlsnaes, T. Risse, & B. A. Simmons (Eds.), *Handbook of International Relations* (pp. 379–400). London: SAGE.

Odell, J. S., & Sell, S. K. (2006). Reframing the Issue: The WTO Coalition on Intellectual Property and Public Health, 2001. In J. S. Odell (Ed.), *Negotiating Trade: Developing Countries in the WTO and NAFTA* (pp. 85–114). Cambridge: Cambridge University Press.

Panke, D. (2012). Dwarfs in international negotiations: How small states make their voices heard. *Cambridge Review of International Affairs, 25*(3), 313–328.

Rubin, J. Z., & Zartman, I. W. (2000). *Power and Negotiation*. Ann Arbor: University of Michigan Press.

Sutter, J. D. (2018, December 14). Can a coalition of 'superheroes' save the COP24 climate talks? Retrieved from https://edition.cnn.com/2018/12/13/health/sutter-high-ambition-cop24-climate/index.html

Vihma, A., Mulugetta, Y., & Karlsson-Vinkhuyzen, S. (2011). Negotiating solidarity? The G77 through the prism of climate change negotiations. *Global Change, Peace & Security, 23*(3), 315–334.

Watts, J. (2020). AILAC and ALBA: Differing Visions of Latin America in Climate Change Negotiations. In C. Klöck, P. Castro, F. Weiler, & L. Ø. Blaxekjær (Eds.), *Coalitions in the Climate Negotiations*. Abingdon: Routledge.

Weiler, F., & Castro, P. (2020). "Necessity Has Made Us Allies": The Role of Coalitions in the Climate Change Negotiations. In C. Klöck, P. Castro, F. Weiler, & L. Ø. Blaxekjær (Eds.), *Coalitions in the Climate Change Negotiations*. Abingdon: Routledge.

Woods, B. A., & Kristófersson, D. M. (2016). The state of coalitions in international climate change negotiations and implications for global climate policy. *International Journal of Environmental Policy and Decision Making, 2*(1), 41–68.

Wu, J., & Thill, J.-C. (2018). Climate change coalition formation and equilibrium strategies in mitigation games in the post-Kyoto Era. *International Environmental Agreements: Politics, Law and Economics, 18*(4), 573–598.

Yamin, F., & Depledge, J. (2004). *The International Climate Change Regime: A Guide to Rules, Institutions and Procedures*. Cambridge: Cambridge University Press.

Appendix 1: Countries and their coalition memberships in the climate negotiations

Carola Klöck

The following table summarises in which coalition(s), if any, each Party to the UNFCCC (except the EU) participates. Data as of March 2020.

Country	ABU	AILAC	ALBA	AOSIS	AGN	Arab Group	BASIC	CACAM	CARICOM	Cartagena	CfRN	COMIFAC	Congo Basin	CVF/V20	EU	EIG	G77	HAC	LDCs	LMDCs	MLDCs	OPEC	P-SIDS	SICA	Umbrella	Sum
Afghanistan														✓			✓	✓	✓		✓					5
Albania								✓																		1
Algeria					✓	✓											✓			✓		✓				5
Andorra																										0
Angola					✓												✓		✓	✓		✓				5
Antigua and Barbuda	✓			✓					✓								✓	✓								5
Argentina										✓	✓						✓	✓								4
Armenia								✓													✓					2
Australia										✓								✓							✓	3
Austria															✓			✓								2
Azerbaijan								✓																		1
Bahamas				✓					✓								✓	✓								4
Bahrain						✓											✓									2
Bangladesh										✓	✓			✓			✓	✓	✓							6
Barbados				✓					✓	✓				✓			✓	✓								6
Belarus																									✓	1
Belgium															✓			✓								2
Belize				✓					✓		✓						✓	✓						✓		6
Benin					✓												✓	✓	✓							4
Bhutan														✓			✓	✓	✓							4
Bolivia			✓														✓			✓						3
Bosnia and Herzegovina											✓															1
Botswana					✓												✓	✓								3

Country												Total
Brazil												4
Brunei Darussalam												1
Bulgaria												2
Burkina Faso												5
Burundi												6
Cambodia												5
Cameroon												5
Canada												2
Cape Verde												4
Central African Republic												7
Chad												5
Chile												3
China												3
Colombia												5
Comoros												7
Congo (DRC)												9
Congo (Rep.)												6
Cook Islands												2
Costa Rica												6
Cote d'Ivoire												2
Croatia												2
Cuba												5
Cyprus												2
Czech Republic												2
Denmark												3
Djibouti												5
Dominica												7
Dominican Republic												7
Ecuador												4
Egypt												4

(Continued)

(Continued)

Country	ABU	AILAC	ALBA	AOSIS	AGN	Arab Group	BASIC	CACAM	CARICOM	Cartagena	CfRN	COMIFAC	Congo Basin	CVF/V20	EU	EIG	G77	HAC	LDCs	LMDCs	MLDCs	OPEC	P-SIDS	SICA	Umbrella	Sum
El Salvador																	✓	✓						✓		3
Equatorial Guinea					✓						✓	✓	✓				✓			✓		✓				7
Eritrea					✓												✓	✓	✓							4
Estonia															✓			✓								2
Eswatini					✓												✓	✓								3
Ethiopia					✓					✓				✓			✓	✓	✓							6
Fiji				✓							✓			✓			✓	✓					✓			6
Finland															✓			✓								2
France										✓					✓			✓								3
Gabon					✓						✓	✓	✓				✓					✓				6
Gambia					✓					✓				✓			✓	✓	✓							6
Georgia																✓		✓								2
Germany										✓					✓			✓								3
Ghana					✓					✓	✓			✓			✓									5
Greece															✓			✓								2
Grenada				✓					✓	✓				✓			✓	✓								6
Guatemala		✓								✓				✓			✓	✓						✓		6
Guinea					✓												✓		✓							3
Guinea-Bissau				✓	✓												✓	✓	✓							5
Guyana				✓					✓		✓						✓	✓								5
Haiti				✓					✓					✓			✓	✓	✓							6
Honduras		✓									✓			✓			✓							✓		5

Hungary
Iceland
India
Indonesia
Iran
Iraq
Ireland
Israel
Italy
Jamaica
Japan
Jordan
Kazakhstan
Kenya
Kiribati
Korea (North)
Korea (South)
Kuwait
Kyrgyz Republic
Lao PDR
Latvia
Lebanon
Lesotho
Liberia
Libya
Liechtenstein
Lithuania
Luxembourg
Madagascar
Malawi

(Continued)

(Continued)

Country	ABU	AILAC	ALBA	AOSIS	AGN	Arab Group	BASIC	CACAM	CARICOM	Cartagena	CfRN	COMIFAC	Congo Basin	CVF/V20	EU	EIG	G77	HAC	LDCs	LMDCs	MDCs	OPEC	P-SIDS	SICA	Umbrella	Sum
Malaysia											✓						✓			✓						3
Maldives				✓							✓			✓			✓	✓								5
Mali					✓					✓				✓			✓	✓	✓							6
Malta															✓			✓								2
Marshall Islands				✓						✓				✓			✓	✓					✓			6
Mauritania					✓	✓											✓	✓	✓							5
Mauritius				✓	✓												✓	✓								4
Mexico								✓		✓						✓										3
Micronesia (Fed. States)				✓													✓	✓					✓			4
Moldova																	✓									1
Monaco																✓										1
Mongolia														✓			✓									2
Montenegro																										0
Morocco					✓	✓											✓	✓								4
Mozambique					✓						✓						✓	✓	✓							5
Myanmar																	✓	✓	✓							3
Namibia					✓												✓	✓								3
Nauru				✓													✓	✓					✓			4
Nepal														✓			✓	✓	✓							4
Netherlands										✓					✓			✓								3
New Zealand										✓															✓	2
Nicaragua			✓											✓			✓			✓				✓		5
Niger					✓									✓			✓	✓	✓							5

Country		Count
Nigeria	✓ ✓	4
Niue	✓ ✓	2
Norway	✓ ✓	3
Oman	✓ ✓	2
Pakistan	✓	3
Palau	✓ ✓	4
Palestine	✓ ✓	3
Panama	✓ ✓ ✓	5
Papua New Guinea	✓ ✓ ✓ ✓	6
Paraguay	✓ ✓ ✓	3
Peru	✓ ✓ ✓	3
Philippines	✓	2
Poland	✓	2
Portugal	✓ ✓	2
Qatar	✓ ✓	2
Romania	✓	3
Russia		2
Rwanda	✓ ✓	1
Samoa	✓ ✓	7
San Marino	✓ ✓	7
São Tomé e Príncipe	✓ ✓	0
Saudi Arabia	✓ ✓	5
Senegal	✓ ✓	4
Serbia	✓	5
Seychelles	✓ ✓	0
Sierra Leone	✓ ✓	4
Singapore	✓ ✓	5
Slovakia	✓	4
Slovenia		2
Solomon Islands	✓	2
		6

(Continued)

(Continued)

Country	ABU	AILAC	ALBA	AOSIS	AGN	Arab Group	BASIC	CACAM	CARICOM	Cartagena	CfRN	COMIFAC	Congo Basin	CVF/V20	EU	EIG	G77	HAC	LDCs	LMDCs	MLDCs	OPEC	P-SIDS	SICA	Umbrella	Sum
Somalia					✓	✓											✓	✓	✓							5
South Africa					✓		✓				✓						✓									4
South Sudan					✓												✓		✓	✓						4
Spain										✓					✓			✓								3
Sri Lanka														✓			✓			✓						3
St Kitts and Nevis				✓					✓								✓									3
St Lucia				✓					✓	✓				✓			✓	✓								6
St Vincent and the Grenadines				✓					✓								✓	✓								4
Sudan					✓	✓				✓	✓			✓			✓		✓	✓						8
Suriname				✓					✓	✓	✓						✓									5
Sweden										✓					✓			✓								3
Switzerland																✓		✓								2
Syria						✓											✓			✓						3
Tajikistan								✓									✓			✓	✓					4
Tanzania					✓					✓	✓			✓			✓		✓							6
Thailand										✓				✓			✓									3
Timor-Leste				✓										✓			✓		✓	✓						5
Togo					✓						✓						✓		✓							4
Tonga				✓													✓						✓			3
Trinidad and Tobago				✓					✓					✓			✓									4
Tunisia					✓	✓								✓			✓									4
Turkey																										0

Country											Count
Turkmenistan									✓		2
Tuvalu			✓	✓	✓				✓	✓	5
Uganda	✓		✓	✓	✓					✓	6
Ukraine											1
United Arab Emirates		✓	✓	✓	✓						4
United Kingdom											2
United States											2
Uruguay		✓	✓	✓					✓		3
Uzbekistan											1
Vanuatu	✓	✓	✓	✓	✓		✓	✓	✓		7
Venezuela		✓	✓	✓	✓					✓	4
Vietnam			✓	✓	✓						4
Yemen		✓	✓	✓	✓					✓	5
Zambia		✓	✓	✓	✓					✓	5
Zimbabwe											2

Appendix 2: Coalitions in the climate negotiations

Carola Klöck, Florian Weiler, and Paula Castro

There are around 25 different coalitions that are, or were, active in the climate change negotiations. The precise number of coalitions will depend on the specific definition used; here we focus on groups of countries that have coordinated their positions and made joint statements during the negotiations for more than just one COP (Castro & Klöck, 2020; Klöck, Weiler, Castro, & Blaxekjær, 2020). We briefly describe each coalition and outline: when the coalition became active; what its main positions or objectives are; and who the members are.

ABU (Argentina-Brazil-Uruguay)

Short description: Argentina-Brazil-Uruguay (ABU) is a very recent regional coalition, formed among a small group of countries with very strong historical, economic, and political ties, not the least including common membership of the Mercosur trade bloc. The three countries started informally coordinating common positions, initially on adaptation, around the negotiations on the Paris Agreement (Lorenzo Arana, 2020), and have since started identifying as an actual coalition, while becoming active across a wider range of issues. Since the three countries are important agricultural producers, they also contribute to discussions on mitigation and adaptation in the agricultural sector (Moosmann, Urrutia, Siemons, Cames, & Schneider, 2019).
   Members: Argentina, Brazil, Uruguay.
   Website: ABU does not have a website.

AGN (African Group of Climate Change Negotiators)

Short description: The African Group of Climate Change Negotiators (AGN) is the only regional UN group that is also active as a substantive negotiation coalition. Formally, African participation has three tiers. The Conference of African Heads of State and Government on Climate Change is the highest political tier and meets during Summits of the African Union. The African Ministerial Conference on the Environment provides political guidance and mandates the AGN. The AGN is the technical tier of the African climate negotiation structure and engages in the UNFCCC process (AGN, 2019).

Covering the 54 countries located in Africa, the AGN has long been comparatively silent and sidelined, not least because the countries are very diverse and united not by common interests, but by geographical location. However, since the 2009 Copenhagen Summit (COP15), the AGN has become more active and sought to present a more united front (Chin-Yee, Nielsen, & Blaxekjær, 2020; Gupta, 2014; Roger, 2013; Roger & Belliethathan, 2016; Tsega, 2016). Key points for AGN are the recognition of Africa as "particularly vulnerable", as well as financial and technological support for mitigation as well as, more importantly, adaptation efforts.

Members: Algeria, Angola, Benin, Botswana, Burkina Faso, Burundi, Cameroon, Cabo Verde, Central African Republic, Chad, Comoros, Democratic Republic of the Congo, Republic of the Congo, Côte d'Ivoire, Djibouti, Egypt, Equatorial Guinea, Eritrea, Eswatini, Ethiopia, Gabon, Gambia, Ghana, Guinea, Guinea-Bissau, Kenya, Lesotho, Liberia, Libya, Madagascar, Malawi, Mali, Mauritania, Mauritius, Morocco, Mozambique, Namibia, Niger, Nigeria, Rwanda, São Tomé and Príncipe, Senegal, Seychelles, Sierra Leone, Somalia, South Africa, South Sudan, Sudan, Tanzania, Togo, Tunisia, Uganda, Zambia, Zimbabwe.

Website: https://africangroupofnegotiators.org/about-the-agn/.

AILAC (Independent Association of Latin America and the Caribbean)

Short description: At the 2012 COP18 in Doha, six countries (Chile, Colombia, Costa Rica, Guatemala, Panama, and Peru) formally created the Independent Association of Latin America and Caribbean States (Asociación Independiente de Latinoamérica y Caribe, AILAC), but these countries had already been coordinating their participation in climate negotiations before that. In 2015, Honduras and Paraguay joined AILAC.

AILAC is partly a reaction to ALBA, and takes a conciliatory, "middle ground" approach. It wants to build bridges between different groups and positions, and notably across the divide between developing and developed countries. By enhancing trust in the negotiations AILAC seeks to create room for compromise and consensus to move the negotiations forward (Edwards & Roberts, 2015, 2016; Watts, 2020; Watts & Depledge, 2018).

AILAC seeks ambitious climate policies. While developed countries should take the lead and provide financial assistance, according to AILAC, the group also wants to move the strict divide between developed and developing countries in terms of mitigation commitments. As opposed to the traditional G77 position, it argues that the principle of Common but Differentiated Responsibilities and Respective Capabilities (CBDR-RC) does not mean that developing countries are not to adopt emission reduction targets (Watts, 2020). Developing countries should not wait for developed countries to lead, but should embark on a low-carbon transition themselves (Ciplet & Roberts, 2019).

Members: Chile, Colombia, Costa Rica, Guatemala, Honduras, Panama, Paraguay, and Peru.

Website: http://ailac.org/en/sobre/.

ALBA (Bolivarian Alliance for the Peoples of Our America)

Short description: The Bolivarian Alliance for the Peoples of Our America (Alianza Bolivariana de los Pueblos de Nuestra América, ALBA) is the climate arm of the ALBA People's Trade Treaty (ALBA-TCP), a trade treaty covering eight countries in Latin America and the Caribbean. In the climate context, ALBA unites five countries: Bolivia, Cuba, Dominica, Nicaragua, Venezuela.

ALBA emerged as a climate coalition in 2009, at the controversial Copenhagen climate summit (COP15). ALBA has adopted an uncompromising and radical position and criticises the neoliberal US-dominated international order. In the climate context, ALBA members highlight climate injustices and the need for equity, often with fiery rhetoric (Edwards & Roberts, 2015). In particular, they emphasise the historical responsibility – or historical "debt" – of developed countries for climate change, and object to market mechanisms as a "solution" for mitigation (Watts, 2020). While the group was very vocal in Copenhagen, its activities and influence have declined since, not least because of domestic problems in Venezuela (Edwards & Roberts, 2015; Watts, 2020; Watts & Depledge, 2018).

Members: Bolivia, Cuba, Dominica, Nicaragua, Venezuela. Antigua and Barbuda, Grenada, St Kitts and Nevis, St Lucia, and St Vincent and the Grenadines are members of the wider ALBA-TCP project, but not actively engaged in the ALBA coalition in the climate negotiations.

Website: ALBA no longer maintains a website.

Arab Group (League of Arab States)

Short description: The League of Arab States, also referred to as the Arab League or Arab Group in the climate negotiations, is a regional organisation of 22 Arabic countries. In the climate change negotiations, the Arab States of Northern Africa tend to engage more through the African Group of Negotiators (Luomi, 2011). The remaining Arab states are divided: the oil-rich Gulf States, notably Saudi Arabi, as well as, to a lesser degree, Kuwait and Qatar, have played a very active, but mostly negative role in the climate change negotiations, in an effort to protect their fossil-fuel dependent economies from ambitious climate policies (Depledge, 2008; Flisnes, 2019). Dominated by Saudi Arabia, they mostly coordinate through the Gulf Cooperation Council, OPEC, and the Organization of Arab Petroleum Exporting Countries, OAPEC. In contrast, the resource-poorer Arab states have not been very active in the climate change negotiations, and have, implicitly or explicitly, consented to the OPEC position. Accordingly, Luomi (2011, p. 255) concludes that the "functioning and relevance of the League of Arab States as an interest aggregate, despite an all-inclusive membership and increasing attempts at co-ordination, has for the most part been low" (see also Luomi, 2010). Nonetheless, the Arab Group does regularly meet and exchange information during climate summits, even if they rarely speak as a group (Yamin & Depledge, 2004, p. 48).

Members: Algeria, Bahrain, Comoros, Djibouti, Egypt, Iraq, Jordan, Kuwait, Lebanon, Libya, Mauritania, Morocco, Oman, Palestine, Qatar, Saudi Arabia, Somalia, Sudan, Syria, Tunisia, United Arab Emirates, and Yemen.

Website: The Arab Group does not have a website.

AOSIS (Alliance of Small Island States)

Short description: The United Nations recognises about 50 small island developing states (SIDS) and territories worldwide that share a common vulnerability to the adverse effects of climate change. The 39 independent SIDS (as well as five observers) have defended their island interests through AOSIS since the very beginning of climate negotiations, and while the individual AOSIS members are fairly small, and have little economic and political influence, AOSIS as a group was surprisingly successful, although in later periods their success levels somewhat decreased (Betzold, 2010; Betzold, Castro, & Weiler, 2012). Among other things, SIDS have secured a specific seat on the Board of the COP and other bodies, and AOSIS is regularly consulted in smaller groups, the so-called Friends of the Chair. As such, AOSIS is today recognised as a key player in the climate negotiations that actively shapes the debate through regular submissions and proposals. AOSIS was also the first group to submit a full draft protocol during the Kyoto Protocol negotiations, and sought emission reductions of 20% from 1990 levels by 2005 (UNFCCC, 2002). AOSIS generally fights for ambitious mitigation policies to keep the global temperature rise to below 1.5°C. AOSIS has also a strong interest in support for adaptation, as well as loss and damage, and increasingly promotes oceans and the blue economy to address climate change (Carter, 2020).

Members: Antigua and Barbuda, Bahamas, Barbados, Belize, Cabo Verde, Comoros, Cook Islands, Cuba, Dominica, Dominican Republic, Fiji, Grenada, Guinea-Bissau, Guyana, Haïti, Jamaica, Kiribati, Maldives, Marshall Islands, Mauritius, Micronesia (Federated States of), Nauru, Niue, Palau, Papua New Guinea, Samoa, São Tomé and Príncipe, Seychelles, Singapore, Solomon Islands, St Kitts and Nevis, St Lucia, St Vincent and the Grenadines, Suriname, Timor-Leste, Tonga, Trinidad and Tobago, Tuvalu, Vanuatu.

The five observers are American Samoa, Guam, Netherlands Antilles, Puerto Rico, and the United States Virgin Islands.

Website: www.aosis.org/.

BASIC (Brazil-South Africa-India-China)

Short description: Brazil, South Africa, India, and China differ widely in terms of their economic and energy structures, greenhouse gas emissions, and political systems, and therefore do not automatically have common climate interests (Ciplet, Khan, & Roberts, 2015). Yet, these four countries share a common growing status in world affairs (Hallding, Jürisoo, Carson, & Atteridge, 2013), and faced increasing pressure to reduce their own greenhouse gas emissions from both developed

Annex I countries and developing non-Annex I countries, especially SIDS and LDCs. In response, BASIC "began to see more value in working together than in being tethered to their former peers in the splintering G77" (Ciplet et al., 2015, p. 72) and announced a joint approach to the climate negotiations just before the 2009 Copenhagen summit. Alongside the USA, it was BASIC that dominated the Copenhagen COP (Happaerts & Bruyninckx, 2013), even being considered the main "villains" of the meeting (Hurrell & Sengupta, 2012).

BASIC emphasise their developing country identity, distinct from developed countries, and therefore also explicitly understand themselves as a part of G77 (Happaerts & Bruyninckx, 2013; Hochstetler & Milkoreit, 2014). This also means that BASIC insists on differentiation and the principle of "common but differentiated responsibilities and respective capabilities", and hence objects to binding emission reduction targets for themselves, although they are less opposed to *voluntary* commitments. Beyond their opposition to binding emission reductions, differences abound, and tensions emerged within the group directly after Copenhagen. South Africa took a more conciliatory approach (not least as host of COP17 in Durban in 2011), as did Brazil, whereas India and China joined the Like-Minded Developing Countries (Hallding et al., 2013; Hurrell & Sengupta, 2012). Nevertheless, BASIC continue to hold quarterly ministerial meetings on climate change, and to make joint statements in the negotiations. Cooperation does, however, focus on broad principles rather than the finer details of their implementation (Hallding et al., 2013).

Members: Brazil, South Africa, India, and China
Website: BASIC does not have a website.

CACAM (Central Asia, Caucasus, Albania, and Moldova Group)

Short description: The Central Asia, Caucasus, Albania, and Moldova Group (CACAM) was created in 2000. Its seven members – Albania, Armenia, Georgia, Kazakhstan, Moldova, Turkmenistan, and Uzbekistan – are fairly diverse and have not historically enjoyed friendly relationships. However, they all shared a common concern about their specific status within the climate regime. The Convention and the Kyoto Protocol distinguish between Annex I countries – developed countries and countries with economies in transition – and non-Annex I countries – developing countries. CACAM members consider themselves as not falling in either category, as non-Annex I Parties with economies in transition. Accordingly, CACAM activities mainly focused on clarifying their position in the regime, and rarely expressed common positions on other issues (Yamin & Depledge, 2004). Nevertheless, CACAM regularly met during climate summits until 2009, and also helped members to overcome well-known problems of small delegations, under Kazakh leadership (Sabonis-Helf, 2003).

Members: Albania, Armenia, Georgia, Kazakhstan, Moldova, Turkmenistan, and Uzbekistan.
Website: CACAM does not have a website.

CARICOM (Caribbean Community)

Short description: The Caribbean Community (CARICOM) is a regional organisation that was founded in 1973 by Barbados, Guyana, Jamaica, and Trinidad and Tobago. It has since grown to 20 members (of which five are associate members, and of which 14 are parties to the UNFCCC). CARICOM pursued a regional approach to trade negotiations, and has been present at climate summits since 2004. The 5C – the Caribbean Community Climate Change Centre – coordinates the regional response to climate change (but not climate change diplomacy).

All CARICOM members that are party to the UNFCCC are also members of AOSIS, and engage in the negotiations through AOSIS, but CARICOM helps represent Caribbean viewpoints within AOSIS. CARICOM largely shares the same concerns and priorities, namely, ambitious mitigation to ensure global warming remains below 1.5°C, as well as a focus on adaptation, loss and damage, and climate finance (see e.g. CARICOM, 2015).

Members: Antigua and Barbuda, Bahamas, Barbados, Belize, Dominica, Grenada, Guyana, Haïti, Jamaica, St Lucia, St Kitts and Nevis, St Vincent and the Grenadines, Suriname, Trinidad and Tobago.

Montserrat is also a member but not party to the UNFCCC; Anguilla, Bermuda, British Virgin Islands, Cayman Islands, and Turks and Caicos are Associate Members.

Website: https://caricom.org/.

Cartagena Dialogue (for Progressive Action)

Short description: As the name indicates, the Cartagena Dialogue for Progressive Action is an informal platform or *dialogue* rather than a regular negotiation coalition. It seeks to promote progressive climate policies (Lynas, 2011). Cartagena has its roots in a long-standing loose network of experienced negotiators from progressive countries (including some members of AOSIS, LDCs, the EU, and Latin America and the Caribbean). After the failed Copenhagen Summit (COP15), this network was formalised as the Cartagena Dialogue, in a meeting in the Colombian city of Cartagena in 2010 (Blaxekjær, 2016).

The Cartagena Dialogue is best understood as an informal and internal space to bridge the traditional North–South divide. It is open to countries from the Global North and the Global South that want ambitious, comprehensive, and legally binding global climate agreements. Its purpose is not to bargain and push common positions in the negotiations. Rather, it allows parties to appraise and explore each other's points of view, and thus opens up space for compromise and progress (Carter, 2018, p. 140). Through its very flexible, inclusive format that is open to negotiators, but closed to observers and the media, the Cartagena Dialogue helped to create new communication channels, find middle ground, and move forward the UNFCCC process at a decisive and difficult moment (Edwards & Roberts, 2015). It thus was instrumental in bringing the negotiations "back on track" (Blaxekjær, 2020) and laying the groundwork for the 2015 Paris Agreement.

Members: Because of its closed, flexible, and fluid nature, the Cartagena Dialogue has no well-defined membership. Blaxekjær (2020) lists the following as of COP19: Antigua and Barbuda, Australia, Bangladesh, Barbados, Burundi, Chile, Colombia, Costa Rica, Denmark, Dominican Republic, Ethiopia, European Union, France, Gambia, Georgia, Germany, Ghana, Grenada, Guatemala, Indonesia, Kenya, Lebanon, Malawi, Maldives, Marshall Islands, México, Netherlands, New Zealand, Norway, Panama, Peru, Rwanda, Samoa, Spain, Swaziland, Sweden, Switzerland, Tajikistan, Tanzania, Uganda, United Arab Emirates, and the United Kingdom.

Website: The Cartagena Dialogue does not have a website.

CfRN (Coalition for Rainforest Nations)

Short description: The Coalition for Rainforest Nations unites 53 developing countries with tropical rainforest cover. In 2004, the Prime Minister of Papua New Guinea and the President of Costa Rica co-founded the CfRN, in an effort to obtain financial support not only for afforestation, but also for conserving forests (Edwards & Roberts, 2015). The CfRN achieved this objective when COP13 adopted a decision on "reducing emissions from deforestation in developing countries" (Decision 2/CP.13), which has since been expanded into REDD+ (Reducing Emissions from Deforestation and Forest Degradation and the role of conservation, sustainable management of forests and enhancement of forest carbon stocks in developing countries). Under REDD+, developing countries receive "results-based compensation for preventing deforestation and degradation, and for conserving and enhancing carbon stocks" (Halstead, 2018).

The CfRN continues to promote forestry and forest conservation as a key mechanism to address climate change, and to increase financial support for this. Thanks to their advocacy, the Paris Agreement dedicates a full article to forestry (Article 5). Forestry is thus the only sector with its own article in the Paris Agreement (Halstead, 2018).

Members: Argentina, Bangladesh, Belize, Botswana, Cambodia, Cameroon, Central African Republic, Congo (Democratic Republic), Congo (Republic), Costa Rica, Dominica, Dominican Republic, Ecuador, Equatorial Guinea, Fiji, Gabon, Ghana, Guatemala, Guyana, Honduras, India, Indonesia, Jamaica, Kenya, Laos, Lesotho, Liberia, Madagascar, Malawi, Malaysia, Mali, Mozambique, Namibia, Nicaragua, Nigeria, Pakistan, Panama, Papua New Guinea, Paraguay, Samoa, Sierra Leone, Singapore, Solomon Islands, South Africa, St Lucia, Sudan, Suriname, Thailand, Uganda, Uruguay, Vanuatu, Vietnam, and Zambia.

It should be noted that not all CfRN interventions and submissions are supported by all members.[1]

Website: www.rainforestcoalition.org/about/.

CVF (Climate Vulnerable Forum)

Short description: The Climate Vulnerable Forum describes itself as a "partnership of countries highly vulnerable to a warming planet" (https://thecvf.org/

about/). It grew out of a meeting of 11 countries in Malé, Maldives, shortly before the 2009 Copenhagen Summit (COP15), and now unites 48 countries from across the Global South. The Ministers of Finance of some CVF members regularly meet as the V20, the Vulnerable 20, a subgroup of the CVF, to exchange and promote best practices on climate finance and to engage in joint advocacy.

The CVF seeks to raise awareness and ambition. Although their own emissions are fairly low, members committed to domestic low-emission development pathways, but at the same time underline their dependence on ambitious mitigation by large polluters (see Climate Vulnerable Forum, 2016). In this context, the CVF published its own vulnerability indicator, the Climate Vulnerability Monitor, in 2010 and 2012, and released the Low Carbon Monitor in 2016 that examines the benefits and opportunities of limiting global warming to 1.5°C. This 1.5° target is at the core of the CVF, and this coalition has been instrumental in ensuring that the Paris Agreement specifically refers to 1.5°C beyond the 2°C target (Ciplet & Roberts, 2019; Ott, Bauer, Brandi, Mersmann, & Weischer, 2016).

Members: Afghanistan, Bangladesh, Barbados, Bhutan, Burkina Faso, Cambodia, Colombia, Comoros, Congo (Democratic Republic), Costa Rica, Dominican Republic, Ethiopia, Fiji, The Gambia, Ghana, Grenada, Guatemala, Haïti, Honduras, Kenya, Kiribati, Lebanon, Madagascar, Malawi, Maldives, Marshall Islands, Mongolia, Morocco, Nepal, Niger, Palau, Palestine, Papua New Guinea, Philippines, Rwanda, St Lucia, Samoa, Senegal, South Sudan, Sri Lanka, Sudan, Tanzania, Timor-Leste, Tunisia, Tuvalu, Vanuatu, Vietnam and Yemen.

Website: https://thecvf.org/.

COMIFAC (Commission des Forêts d'Afrique Centrale)

Short description: The Central African Forest Commission (Commission des Forêts d'Afrique Centrale, COMIFAC) is a regional organisation in Central Africa that aims at coordinating and promoting the sustainable management of forests in Central Africa (while maintaining the right of their people to rely on forest resources for their economic and social development), as well as at making the Central African voice heard in international forums on forests and the environment. COMIFAC has its roots in the 1999 Yaoundé Declaration, which led to the creation of the Conference of Ministers in Charge of the Central African Forests in 2000. The Conference of Ministers became COMIFAC in 2005 (COMIFAC, 2009).

In 2005, COMIFAC members started to coordinate their participation in the UNFCCC process, when forests and reducing emissions from deforestation entered the climate agenda (Tadoum et al., 2012). COMIFAC works through thematic working groups. The Climate Working Group was created by the Congo Basin countries (all of which are also COMIFAC members) and has grown to include all COMIFAC countries. The working group meets regularly before and after climate summits to share experiences and coordinate positions (e.g. COMIFAC, 2016). COMIFAC also holds daily coordination meetings during climate summits, and organises side events (e.g. COMIFAC, 2016).

Given their important forest resources, COMIFAC countries are mainly interested in questions around forests and REDD+, and promote sustainable forest management as a key tool to mitigate climate change. COMIFAC seeks financial compensation for their preservation of tropical forests, and promotes carbon markets to fund REDD+. Further priorities are climate finance, technology transfer, and capacity-building (Tadoum et al., 2012).

Members: Burundi, Cameroon, Central African Republic, Chad, Congo (Democratic Republic), Congo (Republic), Equatorial Guinea, Gabon, Rwanda, and São Tomé e Príncipe.

Website: www.comifac.org/.

Congo Basin

Short description: The Central African Congo Basin is the second largest contiguous rainforest after the Amazon. The six Congo Basin countries (Cameroon, Central African Republic, both Congos (Democratic Republic and Republic), Equatorial Guinea, and Gabon) are all also members of COMIFAC. The Congo Basin works closely with COMIFAC; both coalitions defend very similar positions and priorities, and share difficulties related to small delegations, limited resources, and weak capacity (Tadoum, 2009).

The Congo Basin countries started to coordinate their positions and actions within the UNFCCC process in 2005. With COMIFAC support, they created a Climate Working Group, which meets regularly to share experiences, build capacity, and prepare common positions and submissions for climate summits and interim sessions. The Working Group has become the voice of Congo Basin countries in the negotiations, and has since expanded to include all COMIFAC member countries (COMIFAC, 2016; Tadoum, 2009). Nevertheless, the Congo Basin countries regularly meet as Congo Basin countries during COPs.

Just like the larger COMIFAC, the Congo Basin countries focus their work on forestry issues and REDD+, and the role of forests in climate change mitigation. Furthermore, as African countries, they put special emphasis on financial assistance, capacity-building, and technology transfer (Tadoum, 2009).

Members: Cameroon, Central African Republic, Congo (Democratic Republic), Congo (Republic), Equatorial Guinea, and Gabon.

Website: The Congo Basin does not have a website.

EIG (Environmental Integrity Group)

Short description: The Environmental Integrity Group emerged after the Kyoto Protocol negotiations, and was established in the year 2000 at the initiative of Switzerland. Switzerland originally participated in the climate change negotiations as part of JUSSCANNZ, but did not share the positions of the Umbrella Group that emerged out of JUSSCANNZ (see below). Since the negotiations, and in particular consultations in smaller groups like the "Friends of the Chair", build on coalitions (see Klöck et al., 2020), Switzerland was left out of discussions. In response, it formed a new group in 2000, together with Mexico and South Korea, who are non-Annex I countries but at the same time OECD members (Yamin & Depledge, 2004). The group has since been joined by Liechtenstein, Monaco, and Georgia. The group loosely coordinates its positions, and is often led by Switzerland.

The coalition's membership of both Annex I and non-Annex I countries makes the EIG stand out, and is representative of the EIG's conciliatory and constructive approach to the negotiations. With its unique mix of members, and in line with the coalition's name, the EIG emphasises ambitious climate actions in both the Global North and the Global South. Its members pursue relatively ambitious policies domestically, and argue for emerging economies – such as Mexico – to take on emission reductions (Darby, 2015).

Members: Georgia, Liechtenstein, Mexico, Monaco, South Korea, and Switzerland.

Website: The EIG does not have a website.

EU (European Union)

Short description: The European Union (EU) is a particular coalition, since it is itself a party to the UNFCCC. EU member states coordinate very tightly and always speak as the EU, not as individual countries. The EU is thus the most cohesive group in the climate negotiations (Afionis, 2017; Yamin & Depledge, 2004).

The EU and its member states consider themselves as a progressive leader in the climate change negotiations, and have been seen as such by others (Afionis, 2011, 2017). The EU emphasises the need for multilateralism and promotes ambitious, comprehensive, and fair climate agreements (Bäckstrand & Elgström, 2013). Domestically, the EU seeks to lead by example and has reduced its greenhouse gas emissions through market mechanisms and internal burden-sharing, which are key mechanisms to overcome internal differences in domestic emission profiles and levels of climate ambition (Parker, Karlsson, & Hjerpe, 2017). The EU also provides financial and technological support to help others follow the European example.

The EU has successfully moved the UNFCCC process forward at key moments in the negotiations, for instance, after the US retreat from the Kyoto Protocol in 2001. Yet, it increasingly struggled to maintain its leadership role, notably around the 2009 Copenhagen Summit (COP15) (Bäckstrand & Elgström, 2013). Having been sidelined at Copenhagen, the EU changed its strategy, created a separate Directorate-General for Climate Action (DG CLIMA), and renewed its efforts to forge alliances with other progressive forces, such as SIDS. These efforts contributed to the emergence of the High Ambition Coalition and ultimately the successful completion of the Paris Agreement in 2015.

Members: After the United Kingdom leaving the European Union in 2020, the European Union has 27 member states: Austria, Belgium, Bulgaria, Croatia, Cyprus, Czechia, Denmark, Estonia, Finland, France, Germany, Greece, Hungary,

Ireland, Italy, Latvia, Lithuania, Luxembourg, Malta, Netherlands, Poland, Portugal, Romania, Slovakia, Slovenia, Spain, and Sweden.
   Website: https://ec.europa.eu/clima/.

G77 (Group of 77 and China)

Short description: The Group of 77 and China (G77) was established in 1964 at the first United Nations Conference on Trade and Development. While the G77 retained its original name for symbolic and practical reasons, its membership has grown to currently 134 members, making it the largest coalition in the climate change negotiations. Unsurprisingly, G77 members are extremely diverse, ranging from oil-dependent countries like Saudi Arabia to highly vulnerable island states and least developed countries. Accordingly, it has not always been easy, and sometimes impossible, to maintain unity and defend a common position (Ciplet et al., 2015; Hurrell & Sengupta, 2012; Vihma, Mulugetta, & Karlsson-Vinkhuyzen, 2011). Despite this fragmentation, the G77 remains the core coalition for the Global South in the climate negotiations (Chan, 2020; Williams, 2005).

What holds the G77 together is a common developing country identity and solidarity, based on recognition of the inherent inequalities and injustice of the global economic order (Chan, 2020; Ciplet et al., 2015). In the climate context, this translates into a focus on the historical responsibility of the Global North, which must lead the global response to climate change by reducing its own emissions and providing financial and technological support to the Global South. On other agenda items, the G77 struggles to find common ground, which has also contributed to the growing number of subgroups within G77 – although these subgroups consider themselves to be coalitions within the framework of the overall G77 developing country bloc, not least because of the significant weight that the G77 carries due to its size (Chan, 2020; Klöck et al., 2020).

Members: Afghanistan, Algeria, Angola, Antigua and Barbuda, Argentina, Azerbaijan, Bahamas, Bahrain, Bangladesh, Barbados, Belize, Benin, Bhutan, Bolivia, Botswana, Brazil, Brunei Darussalam, Burkina Faso, Burundi, Cabo Verde, Cambodia, Cameroon, Central African Republic, Chad, Chile, China, Colombia, Comoros, Congo (Democratic Republic), Congo (Republic), Costa Rica, Côte d'Ivoire, Cuba, Djibouti, Dominica, Dominican Republic, Ecuador, Egypt, El Salvador, Equatorial Guinea, Eritrea, Eswatini, Ethiopia, Fiji, Gabon, Gambia, Ghana, Grenada, Guatemala, Guinea, Guinea-Bissau, Guyana, Haïti, Honduras, India, Indonesia, Iran, Iraq, Jamaica, Jordan, Kenya, Kiribati, Kuwait, Laos, Lebanon, Lesotho, Liberia, Libya, Madagascar, Malawi, Malaysia, Maldives, Mali, Marshall Islands, Mauritania, Mauritius, Micronesia (Federated States of), Mongolia, Morocco, Mozambique, Myanmar, Namibia, Nauru, Nepal, Nicaragua, Niger, Nigeria, North Korea, Oman, Pakistan, Panama, Papua New Guinea, Palestine, Paraguay, Peru, Philippines, Qatar, Rwanda, St Kitts and Nevis, St. Lucia, St Vincent and the Grenadines, Samoa, São Tomé e Príncipe, Saudi Arabia, Senegal, Seychelles, Sierra Leone, Singapore, Solomon Islands, Somalia,

South Africa, South Sudan, Sri Lanka, Sudan, Suriname, Syria, Tajikistan, Tanzania, Thailand, Timor-Leste, Togo, Tonga, Trinidad and Tobago, Tunisia, Turkmenistan, Uganda, United Arab Emirates, Uruguay, Vanuatu, Venezuela, Vietnam, Yemen, Zambia, and Zimbabwe.

Website: www.g77.org/.

HAC (High Ambition Coalition)

Short description: The High Ambition Coalition (HAC) emerged during the 2015 Paris climate summit. It started as an initiative of the then Marshallese prime minister, Tony deBrum, who in mid-2015 brought together around 15 states from across the globe, including countries from both the Global North and the Global South. This initiative grew to 79 countries who made a joint public appearance during the Paris summit, and were quickly joined by others, including Brazil and the USA (Carter, 2018; Mathiesen & Harvey, 2015). In the end, HAC united over 100, and up to 120, countries – around four key demands:

> a legally binding agreement, a long-term goal on global warming commensurate with science, a review mechanism to assess emissions commitments, a unified system for tracking countries progress on meeting their goals, and eventually, a more ambitious emissions target of 1.5-degree Celsius temperature change.
>
> (Ciplet & Roberts, 2019)

HAC emerged ad-hoc, in the specific context of the Paris Agreement negotiations. However, in line with expectations that the coalition will continue to operate in one form or another (Ott et al., 2016), HAC reappeared at subsequent COPs, including at COP24 in Katowice and at COP25 in Madrid.

Members: HAC membership is ill-defined, yet the following 106 countries can be considered members: Afghanistan, Angola, Antigua and Barbuda, Australia, Austria, Bahamas, Bangladesh, Barbados, Belgium, Belize, Benin, Bhutan, Brazil, Bulgaria, Burkina Faso, Burundi, Cambodia, Canada, Cape Verde, Central African Republic, Chad, Colombia, Comoros, Congo (Democratic Republic), Croatia, Cuba, Cyprus, Czech Republic, Denmark, Djibouti, Dominica, Dominican Republic, Equatorial Guinea, Eritrea, Estonia, Ethiopia, Fiji, Finland, France, Gambia, Germany, Greece, Grenada, Guinea-Bissau, Guyana, Haïti, Hungary, Ireland, Italy, Jamaica, Kiribati, Laos, Latvia, Lesotho, Liberia, Lithuania, Luxembourg, Madagascar, Malawi, Maldives, Mali, Malta, Marshall Islands, Mauritania, Mauritius, Mexico, Micronesia, Mozambique, Myanmar, Nauru, Nepal, Netherlands, Niger, Norway, Palau, Papua New Guinea, Poland, Portugal, Romania, Rwanda, Samoa, Senegal, Seychelles, Sierra Leone, Singapore, Slovakia, Slovenia, Solomon Islands, Somalia, Spain, St Lucia, St Vincent and the Grenadines, Sudan, Suriname, Sweden, Tanzania, Timor-Leste, Togo, Trinidad and Tobago, Tuvalu, Uganda, United Kingdom, United States of America, Vanuatu, Yemen, Zambia.

Website: The HAC does not have a website.

LDCs (Least Developed Countries)

Short description: The United Nations currently recognises 47 Least Developed Countries (LDCs) whose income, human assets index, and economic vulnerability fall under a given threshold. LDCs are very diverse, and LDCs belong to various other coalitions beyond the LDC group (Kameri-Mbote, 2016). Yet, LDCs increasingly recognised that they share a common vulnerability to climate change and a weak capacity to respond to climate change impacts, as well as very low levels of greenhouse gas emissions. Since their concerns, notably adaptation, were sidelined within the larger G77, LDCs started to coordinate their actions through the LDC Group on Climate Change in 2000–2001 (Ciplet et al., 2015; Yamin & Depledge, 2004).

With limited financial and human resources, LDC delegations tend to be small and struggle to participate effectively in the negotiations, although their bargaining capacity has improved over time and the LDC voice has become louder (Ciplet et al., 2015). Given their own low emissions and high vulnerability, LDCs push for ambitious mitigation policies from larger emitters, and call for substantial support for adaptation, technology transfer, and capacity-building.

Members: Afghanistan, Angola, Bangladesh, Benin, Bhutan, Burkina Faso, Burundi, Cambodia, Central African Republic, Chad, Comoros, Congo (Democratic Republic), Djibouti, Ethiopia, Eritrea, Gambia, Guinea, Guinea-Bissau, Haïti, Kiribati, Laos, Lesotho, Liberia, Madagascar, Malawi, Mali, Mauritania, Mozambique, Myanmar, Nepal, Niger, Rwanda, São Tomé e Príncipe, Senegal, Sierra Leone, Solomon Islands, Somalia, South Sudan, Sudan, Tanzania, Timor-Leste, Togo, Tuvalu, Uganda, Vanuatu, Yemen, and Zambia.

Botswana, Cabo Verde, Maldives, Samoa, and Equatorial Guinea used to be classified as LDCs, but graduated from the category.

Website: www.ldc-climate.org/.

LMDCs (Like-Minded Developing Countries)

Short description: The Like-Minded Developing Countries (LMDCs) group held its first meeting in 2012. While its membership is somewhat fluid and varies over time, it brings together a number of emerging economies such as China and India, but also Malaysia or Thailand, oil exporting countries, such as Saudi Arabia and Venezuela, as well as countries with radical positions such as Bolivia and Ecuador. Together, the LMDCs thus command considerable economic and political clout and account for substantial emissions (Bidwai, 2014).

The LMDCs consider themselves as the core of the developing country group, the G77. It wants to represent a strong, and traditional, developing country voice in the negotiations, and upholds the core principle of "common but differentiated responsibilities and respective capabilities" as enshrined in the 1992 Convention (Blaxekjær, Lahn, Nielsen, Green-Weiskel, & Fang, 2020). LMDCs understand this principle as maintaining the traditional divide of developed (Annex I) and developing (non-Annex I) countries, whereby it is the former that must take the lead in reducing emissions, in line with their historical responsibility, and provide finance, technology transfer, and capacity-building to the latter (Blaxekjær et al., 2020; Edwards & Roberts, 2015; Government of India, 2015). This is largely a reaction to increasing calls for greenhouse gas emission targets for emerging economies and other non-Annex I countries, whose contributions to global greenhouse gas emissions have risen significantly since the 1992 Convention.

Members: LMDC membership is rather fluid. The following countries have repeatedly joined LMDC meetings and submissions: Algeria, Argentina, Bolivia, China, Congo (Democratic Republic), Cuba, Dominica, Ecuador, Egypt, El Salvador, India, Indonesia, Iran, Iraq, Jordan, Kuwait, Libya, Malaysia, Mali, Nicaragua, Pakistan, Qatar, Saudi Arabia, Sri Lanka, Sudan, Syria, Thailand, Venezuela, and Vietnam.

Website: The LMDCs do not have a website.

MLDCs (Mountainous Landlocked Developing Countries)

Short description: Armenia, Kyrgyzstan, and Tajikistan started to cooperate in the climate change negotiations as Mountainous Landlocked Developing Countries (MLDCs) in 2010, and were later joined by Afghanistan. The three founding members share a peculiar status in the climate regime, as they are former Soviet republics with economies in transition, and developing countries (similar to CACAM) (Bhandary, 2017).

MLDCs emphasise their specific vulnerability to climate change, as well as issues related to transportation and food security (Gupta, 2014, p. 152f). Indeed, the IPCC recognises the vulnerability of high mountain ecosystems notably due to retreating glaciers and declining snow cover (Hock et al., 2019). The specific topography and being landlocked further constrain the adaptive capacity of these countries. Because of their limited resources and small delegations, MLDCs are not very active in the climate change negotiations, and recognise that even when the four countries join forces, they lack the resources and capacity to engage more fully in the process (Bhandary, 2017).

Members: Afghanistan, Armenia, Kyrgyz Republic, and Tajikistan.

Website: The MLDCs do not have a website.

OPEC (Organisation of Petroleum Exporting Countries)

Short description: The Organisation of Petroleum Exporting Countries is an international organisation of 13 countries. OPEC members are very diverse, but they tend to strongly depend on oil revenues, especially the Gulf countries. Accordingly, OPEC members fear the adverse economic (and political) effects of ambitious climate policies more than the adverse effects of climate change. Nevertheless, OPEC countries – including the Gulf countries – are vulnerable to climate change, notably in terms of food and water security.

In the climate change negotiations, OPEC is dominated by Saudi Arabia (e.g. Luomi, 2009). Saudi elites fear the political ramifications of a global shift away from oil. Accordingly, Saudi Arabia has sent skilful and experienced negotiators to the various negotiation rounds, who have been very active in the UNFCCC process, in order to slow down negotiations and water down agreements (Depledge, 2008). Other OPEC countries have been much more passive and typically consent to the Saudi position. Saudi Arabia and OPEC also closely coordinate with other Gulf countries through the Gulf Cooperation Council, and have been very influential within the larger G77, including through bullying, financial assistance to poorer countries, appeals to South–South identity and solidarity, and regular chairmanship of the G77 and working groups within it (Barnett, 2008; Depledge, 2008; Luomi, 2010, 2011).

Beyond a general interest in obstructing negotiations and weak climate agreements, OPEC in particular insists on financial compensation for economic losses caused by climate policies. OPEC, and Saudi Arabia in particular, has also repeatedly questioned climate science and emphasised scientific uncertainties, and favours carbon capture and storage (Barnett, 2008; Luomi, 2011).

Members: Algeria, Angola, Congo (Republic), Equatorial Guinea, Gabon, Iran, Iraq, Kuwait, Libya, Nigeria, Saudi Arabia, United Arab Emirates, and Venezuela.

Website: www.opec.org/opec_web/en/.

PSIDS (Pacific Small Island Developing States)

Short description: The Pacific small island developing states (Pacific SIDS or PSIDS) are a regional subgroup comprising the small island states of the Pacific region, which has adopted an increasingly pivotal role within AOSIS and has recently emerged as a coalition under the UNFCCC (Carter, 2020). Given their geography – with most infrastructure and settlements concentrated in low-lying coastal areas – relative remoteness and poor infrastructure, the Pacific SIDS view themselves as being particularly vulnerable to climate change and associated sea-level rise, even in comparison to the larger, better resourced islands of the Caribbean (Pacific SIDS, 2009). While the PSIDS had been meeting and coordinating informally for years, they only emerged as a visible coalition around the Paris Agreement negotiations, following calls from the Pacific Islands Forum and the Pacific Islands Development Forum for more active involvement in the climate talks (Carter, 2015, 2020). The PSIDS rely on cooperation with NGOs and other external actors to overcome their negotiation capacity constraints. Nonetheless, they have achieved several successes under the UNFCCC negotiations, including Tuvalu's role in including loss and damage under the Paris Agreement, Fiji's presidency of COP23 in 2017, and the Marshall Islands' leadership in convening HAC at the Paris summit (Carter, 2020).

Members: Cook Islands, Fiji, Kiribati, Marshall Islands, Micronesia (Federated States of), Nauru, Niue, Palau, Papua New Guinea, Samoa, Solomon Islands, Tonga, Tuvalu, Vanuatu.

Website: The PSIDS do not have a website.

SICA (Central American Coordination System)

Short description: The Central American Coordination System (Sistema de Integración Centroamericano in Spanish, SICA) is a regional organisation that was created in 1991. SICA countries are diverse economically, socially, and politically but are all affected by extreme weather events exacerbated by climate change, such as flooding, torrential rains, and tropical storms. In light of these adverse effects, SICA members have – unsuccessfully – sought recognition as being "particularly vulnerable" to climate change, alongside SIDS and LDCs (Argüello Lopez, 2017).

Because of their vulnerability to climate change, adaptation is a priority for SICA. SICA also emphasises the need to limit global warming to 1.5°C and calls on developed countries to reduce greenhouse gas emissions in light of their historical responsibility (Granados Solís & Madrigal Ramírez, 2014).

Climate negotiators from SICA member countries meet annually and identify a common position ahead of the COP. However, member countries do not use this as the basis for joint negotiations (Granados Solís & Madrigal Ramírez, 2014). SICA struggles to find a common approach to the UNFCCC process, and members tend to engage primarily through the other coalitions in which they are a member, such as AOSIS, AILAC, or ALBA, leading to a weak overall SICA presence in climate negotiations (Argüello Lopez, 2017).

Members: Belize, Costa Rica, Dominican Republic, El Salvador, Guatemala, Honduras, Nicaragua, and Panama.

Website: www.sica.int/.

Umbrella Group

Short description: The Umbrella Group has its roots in the so-called JUSSCANNZ group, which stands for Japan, the United States, Switzerland, Canada, Australia, Norway, and New Zealand (although membership has varied over time). The Umbrella Group includes JUSSCANNZ members except for Switzerland, as well as Belarus, Iceland, Israel, Kazakhstan, the Russian Federation, and Ukraine (Yamin & Depledge, 2004). It should be noted that Umbrella Group countries together represent more than 55% of Annex I emissions in 1990; as a result, the Umbrella Group collectively could have blocked the entry into force of the Kyoto Protocol (Sari, 2005).

The Umbrella Group is a very loose group that emphasises solidarity over unity, and "operates to the mantra of 'working together but not tied together'" (Yamin & Depledge, 2004, p. 45). Accordingly, members mostly use the group to share information, but do not necessarily make joint submissions or interventions in the plenary discussions. Instead, key member countries, notably the United States, Australia, Canada, and the Russian Federation, typically take the floor as individual countries, although they do often support each other's statements and positions. The Umbrella Group has a common objective of ensuring cost-effectiveness, flexibility, and the use of market mechanisms to achieve greenhouse gas emission reductions.

Members: Australia, Belarus, Canada, Iceland, Israel, Japan, New Zealand, Kazakhstan, Norway, the Russian Federation, Ukraine, and the United States of America.
Website: The Umbrella Group does not have a website.

Note

1   See for example the CfRN Statement at the Opening Plenary of ADP 2.7 (COP20, Lima, 2 December 2014), which specifically lists the countries having endorsed the statement, available at https://unfccc.int/sites/default/files/adp2-7_opening_statement_by_the_coalition_for_rainforest_nations_02dec2014..pdf.

References

Afionis, S. (2011). The European Union as a negotiator in the international climate change regime. *International Environmental Agreements: Politics, Law and Economics*, *11*, 341–360.

Afionis, S. (2017). *The European Union in International Climate Change Negotiations*. London: Routledge.

AGN. (2019). About the AGN. Retrieved from https://africangroupofnegotiators.org/about-the-agn/

Argüello Lopez, C. M. (2017). *Geopolítica del Cambio Climático, aportes y desafíos para los países que conforman el Sistema de la Integración Centroamericana* (Master thesis). Benemérita Universidad Autonóma de Puebla, Puebla.

Bäckstrand, K., & Elgström, O. (2013). The EU's role in climate change negotiations: From leader to 'leadiator'. *Journal of European Public Policy*, *20*(10), 1369–1386.

Barnett, J. (2008). The worst of friends: OPEC and G-77 in the climate regime. *Global Environmental Politics*, *8*(4), 1–8.

Betzold, C. (2010). 'Borrowing' power to influence international negotiations: AOSIS in the climate change regime, 1990–1997. *Politics*, *30*(3), 131–148.

Betzold, C., Castro, P., & Weiler, F. (2012). AOSIS in the UNFCCC negotiations: From unity to fragmentation? *Climate Policy*, *12*(5), 591–613.

Bhandary, R. R. (2017). Coalition strategies in the climate negotiations: An analysis of mountain-related coalitions. *International Environmental Agreements: Politics, Law and Economics*, *17*(2), 173–190.

Bidwai, P. (2014). The Emerging Economies and Climate Change: A Case Study of the BASIC Grouping. In *Shifting Power: Critical Perspectives on Emerging Economies*. Amsterdam: Transnational Institute.

Blaxekjær, L. Ø. (2016). New Practices and Narratives of Environmental Diplomacy. In G. Sosa-Nunez & E. Atkins (Eds.), *Environment, Climate Change and International Relations*. Bristol: E-International Relations Publishing. (pp. 143–161).

Blaxekjær, L. Ø. (2020). Diplomatic Learning and Trust: How the Cartagena Dialogue Brought UN Climate Negotiations Back on Track. In C. Klöck, P. Castro, F. Weiler, & L. Ø. Blaxekjær (Eds.), *Coalitions in the Climate Change Negotiations*. Abingdon: Routledge.

Blaxekjær, L. Ø., Lahn, B., Nielsen, T. D., Green-Weiskel, L., & Fang, F. (2020). The Narrative Position of the Like-Minded Developing Countries in Global Climate Negotiations. In C. Klöck, P. Castro, F. Weiler, & L. Ø. Blaxekjær (Eds.), *Coalitions in the Climate Negotiations*. Abingdon: Routledge.

CARICOM. (2015). *CARICOM Declaration for Climate Action*. Barbados: Thirty-Sixth Regular Meeting of the Conference of Heads of Government of the Caribbean Community.

Carter, G. (2015). Establishing a Pacific Voice in the Climate Change Negotiations. In G. Fry & S. Tarte (Eds.), *The New Pacific Diplomacy* (pp. 205–220). Canberra: ANU Press.

Carter, G. (2018). *Multilateral consensus decision making: How Pacific island states build and reach consensus in climate change negotiations* (Doctor of Philosophy). The Australian National University, Canberra.

Carter, G. (2020). Pacific Island States and 30 Years of Global Climate Change Negotiations. In C. Klöck, P. Castro, F. Weiler, & L. Ø. Blaxekjær (Eds.), *Coalitions in the Climate Negotiations*. Abingdon: Routledge.

Castro, P., & Klöck, C. (2020). Fragmentation in the Climate Change Negotiations: Taking Stock of the Evolving Coalition Dynamics. In C. Klöck, P. Castro, F. Weiler, & L. Ø. Blaxekjær (Eds.), *Coalitions in the Climate Change Negotiations*. Abingdon: Routledge.

Chan, N. (2020). The Temporal Emergence of Developing Country Coalitions. In C. Klöck, P. Castro, F. Weiler, & L. Ø. Blaxekjær (Eds.), *Coalitions in the Climate Change Negotiations*. Abingdon: Routledge.

Chin-Yee, S., Nielsen, T. D., & Blaxekjær, L. Ø. (2020). One Voice, One Africa: The African Group of Negotiators—Regional Drivers of Climate Policy. In C. Klöck, P. Castro, F. Weiler, & L. Ø. Blaxekjær (Eds.), *Coalitions in the Climate Negotiations*. Abingdon: Routledge.

Ciplet, D., Khan, M., & Roberts, J. T. (2015). *Power in a Warming World: The New Global Politics of Climate Change and the Remaking of Environmental Inequality*. Boston: MIT Press.

Ciplet, D., & Roberts, J. T. (2019). Splintering South: Ecologically Unequal Exchange Theory in a Fragmented Global Climate. In R. Frey, P. Gellert, & H. Dahms (Eds.), *Ecologically Unequal Exchange* (pp. 273–305). Cham: Palgrave Macmillan.

Climate Vulnerable Forum. (2016). Annex 6: CVF vision. Marrakech High Level Meeting. Retrieved from https://www.thecvf.org/wp-content/uploads/2016/11/CVF-Vision-For-Adoption.pdf

COMIFAC. (2009). *Rapport annuel 2009*. Yaoundé: Commission des Forêts d'Afrique Centrale.

COMIFAC. (2016). *Rapport annuel 2016*. Yaoundé: Commission des Forêts d'Afrique Centrale.

Darby, M. (2015). Meet the unlikely climate allies bridging divides in UN talks. Retrieved from https://www.climatechangenews.com/2015/01/20/meet-the-unlikely-climate-allies-bridging-divides-in-un-talks/

Depledge, J. (2008). Striving for no: Saudi Arabia in the climate change regime. *Global Environmental Politics*, 8(4), 9–35.

Edwards, G., & Roberts, J. T. (2015). *Fragmented Continent: Latin America and the Global Politics of Climate Change*. Cambridge, MA: MIT Press.

Edwards, G., & Roberts, J. T. (2016). A new Latin American climate negotiating group: The greenest shoots in the Doha Desert. *Up Front (Brookings Blog)*. Retrieved from https://www.brookings.edu/blog/up-front/2012/12/12/a-new-latin-american-climate-negotiating-group-the-greenest-shoots-in-the-doha-desert/

Flisnes, M. K. (2019). Where You Stand Depends on What You Sell–Saudi Arabia's Obstructionism in the UNFCCC 2012–2018. *CICERO Report*.

Government of India. (2015). Meeting of Negotiators of Like-Minded Developing Countries Concludes; Javadekar Lauds Work Done by LMDC. [Press release] New Delhi: Press Information Bureau, Ministry of Environment, Forest and Climate Change.

Granados Solís, A., & Madrigal Ramírez, R. (2014). Posiciones del Estado de Costa Rica ante las Conferencias de las Partes de la Convención Marco de las Naciones Unidas sobre el Cambio Climático (COPs). In *Análisis 1/2014*. San José: Friedrich Ebert Stiftung Costa Rica.

Gupta, J. (2014). *The History of Global Climate Governance*. Cambridge: Cambridge University Press.

Hallding, K., Jürisoo, M., Carson, M., & Atteridge, A. (2013). Rising powers: The evolving role of BASIC countries. *Climate Policy, 13*(5), 608–631.

Halstead, J. (2018). Climate change cause area report. Retrieved from https://www.rainforestcoalition.org/wp-content/uploads/2019/10/Founders-Pledge-Climate-Change-Report.pdf

Happaerts, S., & Bruyninckx, H. (2013). Rising Powers in Global Climate Governance: Negotiating in the New World Order. In Working Paper N° 124. Leuven: Leuven Centre for Global Governance Studies.

Hochstetler, K., & Milkoreit, M. (2014). Emerging powers in the climate negotiations: Shifting identity conceptions. *Political Research Quarterly, 67*(1), 224–235.

Hock, R., Rasul, G., Adler, C., Cáceres, B., Gruber, S., Hirabayashi, Y., … Steltzer, H. (2019). High Mountain Areas. In H.-O. Pörtner, D. C. Roberts, V. Masson-Delmotte, P. Zhai, M. Tignor, E. Poloczanska, K. Mintenbeck, A. Alegría, M. Nicolai, A. Okem, J. Petzold, B. Rama, & N. M. Weyer (Eds.), *IPCC Special Report on the Ocean and Cryosphere in a Changing Climate*. Cambridge: Cambridge University Press.

Hurrell, A., & Sengupta, S. (2012). Emerging powers, North—South relations and global climate politics. *International Affairs, 88*(3), 463–484.

Kameri-Mbote, P. (2016). The Least Developed Countries and Climate Change Law. In K. R. Gray, R. Tarasofsky, & C. Carlarne (Eds.), *The Oxford Handbook of International Climate Change Law* (pp. 740–760). Oxford: Oxford University Press.

Klöck, C., Weiler, F., Castro, P., & Blaxekjær, L. Ø. (2020). Introduction. In C. Klöck, P. Castro, F. Weiler, & L. Ø. Blaxekjær (Eds.), *Coalitions in the Climate Negotiations*. Abingdon: Routledge.

Lorenzo Arana, I. (2020). Argentina, Brazil, and Uruguay (A-B-U). In M. Bueno Rubial & S. Linda (Eds.), *Negotiating Climate Change Adaptation*. Cham: Springer.

Luomi, M. (2009). Bargaining in the Saudi Bazaar: Common Ground for a Post-2012 Climate Agreement? In Briefing Paper 48. Helsinki: The Finnish Institute of International Affairs.

Luomi, M. (2010). Oil or Climate Politics? Avoiding a Destabilising Resource Split in the Arab Middle East. In Briefing Paper 58. Helsinki: The Finnish Institute of International Affairs.

Luomi, M. (2011). Gulf of interest: Why oil still dominates Middle Eastern climate politics. *Journal of Arabian Studies, 1*(2), 249–266.

Lynas, M. (2011). Thirty 'Cartagena Dialogue' countries work to bridge Kyoto Gap. Retrieved from http://www.marklynas.org/2011/03/thirty-cartagena-dialogue-countries-work-to-bridge-kyoto-gap/

Mathiesen, K., & Harvey, F. (2015, December 8). Climate coalition breaks cover in Paris to push for binding and ambitious deal. *The Guardian*. Retrieved from https://www.the

guardian.com/environment/2015/dec/08/coalition-paris-push-for-binding-ambitious-climate-change-deal

Moosmann, L., Urrutia, C., Siemons, A., Cames, M., & Schneider, L. (2019). *International Climate Negotiations: Issues at Stake in View of the COP25 UN Climate Change Conference in Madrid*. Luxembourg: European Parliament.

Ott, H. E., Bauer, S., Brandi, C., Mersmann, F., & Weischer, L. (2016). *Climate Alliances après Paris: The Potential of Pioneer Climate Alliances to Contribute to Stronger Mitigation and Transformation*. available at https://hermann-e-ott.de/cms/wp-content/uploads/2016/11/PACA_paper_2016_WI-1.pdf.

Pacific SIDS. (2009). *Views on the Possible Security Implications of Climate Change to Be Included in the Report of the Secretary-General to the 64th Session of the United Nations General Assembly*. New York: Permanent Mission of the Republic of Nauru to the United Nations.

Parker, C. F., Karlsson, C., & Hjerpe, M. (2017). Assessing the European Union's global climate change leadership: From Copenhagen to the Paris Agreement. *Journal of European Integration, 39*(2), 239–252.

Roger, C. (2013). African Enfranchisement in Global Climate Change Negotiations. In *Africa Portal Backgrounder* (Vol. 57, pp. 1–10). Waterloo, Canada: The Centre for International Governance Innovation.

Roger, C., & Belliethathan, S. (2016). Africa in the global climate change negotiations. *International Environmental Agreements: Politics, Law and Economics, 16*(1), 91–108.

Sabonis-Helf. (2003). Catching air? Climate change policy in Russia, Ukraine and Kazakhstan. *Climate Policy, 3*(2), 159–170.

Sari, A. (2005). Developing Country Participation: The Kyoto-Marrakech Politics. In HWWA Discussion Paper 33. Hamburg: Hamburgisches Welt-Wirtschafts-Archiv.

Tadoum, M. (2009). Positionnement des Pays de la COMIFAC aux négociations internationales sur les changements climatiques: Atouts et faiblesses. In *Keynote presented at the Forest Day, 10 November 2009*. Yaoundé Centre for International Forest Research.

Tadoum, M., Makonga, V. K. S., Boundzanga, G. C., Bouyer, O., Hamel, O., & Creighton, G. K. (2012). International Negotiations on the Future Climate Regime Beyond 2012: Achievements from Copenhagen to Cancún and Benefits to the Forests of the Congo Basin. In C. de Wasseige, P. de Marcken, N. Bayol, F. Hiol, P. Mayaux, B. Desclée, R. Nasi, A. Billand, P. Defourny, & R. Eba'a Atyi (Eds.), *The Forests of the Congo Basin: State of the Forest 2010* (pp. 157–170). Luxembourg: Publications Office of the European Union.

Tsega, A. H. (2016). Africa in global climate change governance: Analyzing its position and challenges. *International Journal of African Development, 4*(1), 5–18.

UNFCCC. (2002). *A Guide to the Climate Change Convention Process* (Preliminary 2nd edition). Bonn.

Vihma, A., Mulugetta, Y., & Karlsson-Vinkhuyzen, S. (2011). Negotiating solidarity? The G77 through the prism of climate change negotiations. *Global Change, Peace & Security, 23*(3), 315–334.

Watts, J. (2020). AILAC and ALBA: Differing Visions of Latin America in Climate Change Negotiations. In C. Klöck, P. Castro, F. Weiler, & L. Ø. Blaxekjær (Eds.), *Coalitions in the Climate Negotiations*. Abingdon: Routledge.

Watts, J., & Depledge, J. (2018). Latin America in the climate change negotiations: Exploring the AILAC and ALBA coalitions. *Wiley Interdisciplinary Reviews: Climate Change, 9*(6), e533.

Williams, M. (2005). The third world and global environmental negotiations: Interests, institutions and ideas. *Global Environmental Politics, 5*(3), 48–69.

Yamin, F., & Depledge, J. (2004). *The International Climate Change Regime: A Guide to Rules, Institutions and Procedures.* Cambridge: Cambridge University Press.

Index

Abbott, Tony 100
Adarve, Isabel C. 65
Adler, Emanuel 96–97
Advocacy Coalition Framework (AFC) 37, 38–39
Africa Group of the Whole (AGW) 138
African Climate Policy Centre (ACPC) 140
African Development Bank (AfDB) 139, 140, 145
African Group of Negotiators (AGN): CBDR-RC principle incorporation 146–147, 150; CDM, access limitations 143; coalition alignment and priorities 143–144, *144*, 149, 179, 181, 197; coordination and organization 139–140, *140–141*, 196; description and members 196–197; Durban Platform frustrations 142; G77 alliance as subgroup 60, 138, 149; group formation pathways 137–138; Nairobi COP12, turning point 139; negotiation networks, UNFCCC study 41, 42, 44, *45*; "One Voice, One Africa" narrative 142, 145–148, 150; parallel narrative 148–150, 181; Paris COP21 commitment 143–144, 150; proactive individuals 144–145; representation, negotiating strategies 138–139, 142, 147–148, 149, 150; research aims and methodology 136–137; research data sources 137, **151–152**; responsibilities, climate and energy governance 146–147, 149–150; UN regional group 20, 138; vulnerability, climatic and economic 145–146
African Union (AU) 139
Africa Progress Panel 147
Alliance of Small Island States (AOSIS): ALBA member links 160; Berlin

Mandate, G77 disagreements 57; case studies 17; Chair, selection and powers 78, 79; climate change INC proposals 76–77; coordination and bargaining strategy 37, 76–78, 83–84, 180, 199; description and members 199; Durban Platform brokerage 139; emissions reduction commitments 77, 79, 199; formation challenges 58; G77 alliance as subgroup 58–59, 78; geographic scope 20; level of formality 22, 78; loss and damage campaigns 28, 79; member states 88n2; multiple coalition membership benefits 28, 77–78, 160, 181; negotiating bloc 19; negotiation networks, UNFCCC study 41, 42, 44, *45*; NGO support networks 80; objective changes 20, 180; Pacific islands' contribution 73, 83–84, 87; Pacific states leadership 77, 79–80; SIDS recognition, Rio Earth Summit, 1992 76–77; thematic scope 21, 27, 30n2, 39
alliances, contextual terminology 19, 22
Angola 139, 143, 146
Arab Group (League of Arab States) 26, 198–199
Argentina 158
Ashe, John W. 37
Atela, Joanes Odiwuor 28
Audet, René 18–19, 21, 30n2
Australia 100, 142

Bali COP13, 2007: Bali Action Plan 82; REDD programme 82, 202
Barnett, Jon 75–76, 81
BASIC countries: case studies 17; CDM, access misuse 143; Copenhagen COP15 criticisms 92–93, 118, 200; description and members 199–200;

Printed in the United States
by Baker & Taylor Publisher Services

Printed in the United States
By Bookmasters